MAAP would like to thank all our sponsors for 2014

Queensland Government

Australian Government | Australia Council for the Arts

Australian Government
Visual Arts and Craft Strategy
Queensland

Australian Government
Department of Foreign Affairs and Trade

OCAT Shanghai 上海

GRIFFITH UNIVERSITY
ART GALLERY

NAS GALLERY

QUT THE CONFUCIUS INSTITUTE AT QUEENSLAND UNIVERSITY OF TECHNOLOGY

广东美术馆
GUANGDONG MUSEUM OF ART

COPYRIGHTAGENCY CULTURAL FUND

Griffith UNIVERSITY
Queensland College of Art

Australian
Consulate-General
Guangzhou
澳大利亚驻广州总领事馆

Our Korean partners:

Artsonje Center • Lee Hwaik Gallery • ONE & J GALLERY • Opsis Art • Gallery IHN • Gallery SKAPE

IAAP-Media Art Asia Pacific gratefully acknowledges the assistance of the Queensland Government through Arts ueensland. MAAP is supported by the Visual Arts and Craft Strategy, an initiative of the Australian, state and rritory governments. The 2014 program is also supported by the Australian Government through the Australia orea Foundation of the Department of Foreign Affairs and Trade and the Australia Council for the Arts.

Contents

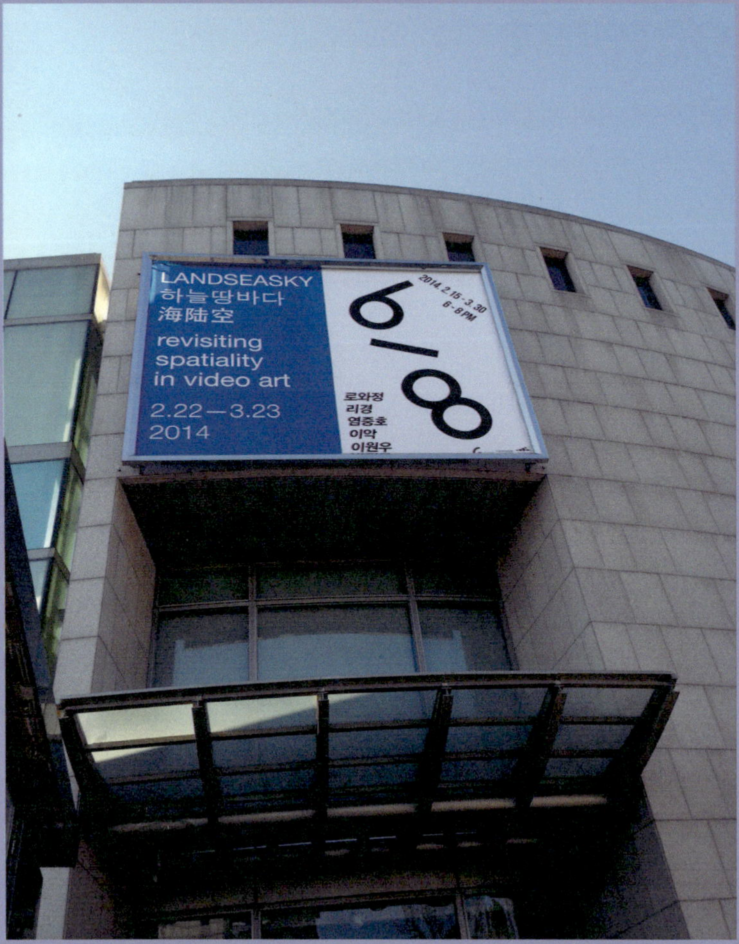

Above, Artsonjie Center building in Seoul; below, the entrance of OCT–Contemporary Art Terminal Shanghai.

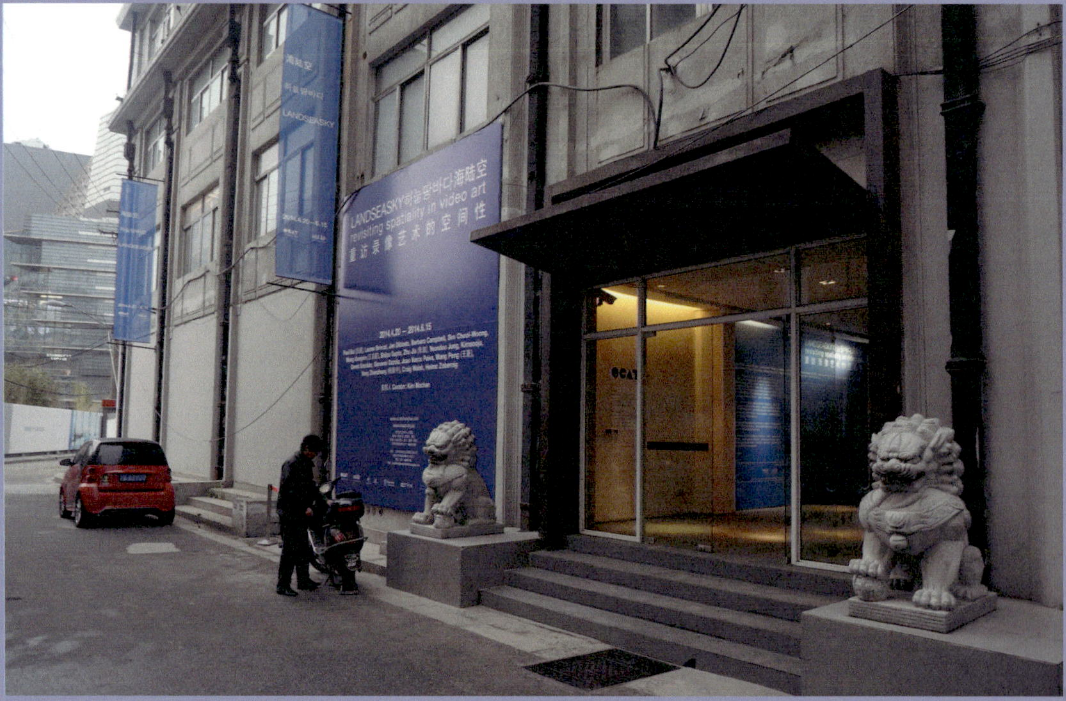

Directors Foreword

2014 was a tremendous year of exhibition delivery both nationally and internationally. I can honestly say it was the most energetic, innovative and productive year that I have ever experienced at MAAP.

2014 is the 17th year of operation since becoming an Incorporated Association in June 1998. It is the eighth year of MAAP Media Bank since we started to note the lending habits that humbly commenced in 2006. 2014 is also the last year of the triennial funding cycle for MAAP SPACE and organisational funding.

The entire year was a whirlwind of activity commencing with the preparation and opening on 22 February of *LANDSEASKY: revisiting spatiality in video art* (LANDSEASKY) in Seoul across six venues. This was a huge success and the massive press – print and television coverage - was outstanding with the exhibition attracting critical praise and a concentrated audience that voiced very positive responses. In Seoul Artsonjc (independent museum) and five commercial galleries presented three artists each in a beautifully balanced trail of exhibition chapters all within 5 minutes walking distance.

Following directly on from Seoul was the exhibition at OCT – Contemporary Art Terminal Shanghai opening on 19 April where the entire exhibition design was built to our specifications in the 1400 square metre 1920's banking warehouse. In China, MAAP also partnered with the National Art Museum of China in Beijing to present five Australian artists work at the internationally acclaimed *Thingworld: International Media Art Triennial* opening in June. Artists' talks, interviews, press conferences and symposiums were realised in both Beijing and Shanghai making a very rich series of events.

To further enhance the exhibition trail a great partnership was struck with Griffith University Art School Gallery in Brisbane and the National Art School Gallery in Sydney. LANDSEASKY opened in Sydney in August and in October as an edition of two sequential exhibition chapters in Brisbane.

It has been an astounding year of exhibitions and this most certainly would not have been possible with the support of our national and international partners, sponsors, our staff, interns along with wonderful volunteers and of course, the MAAP Board that continues to bring excellent advice, strength and stability in these times of flux.

Kim Machan
Director, MAAP - Media Art Asia Pacific

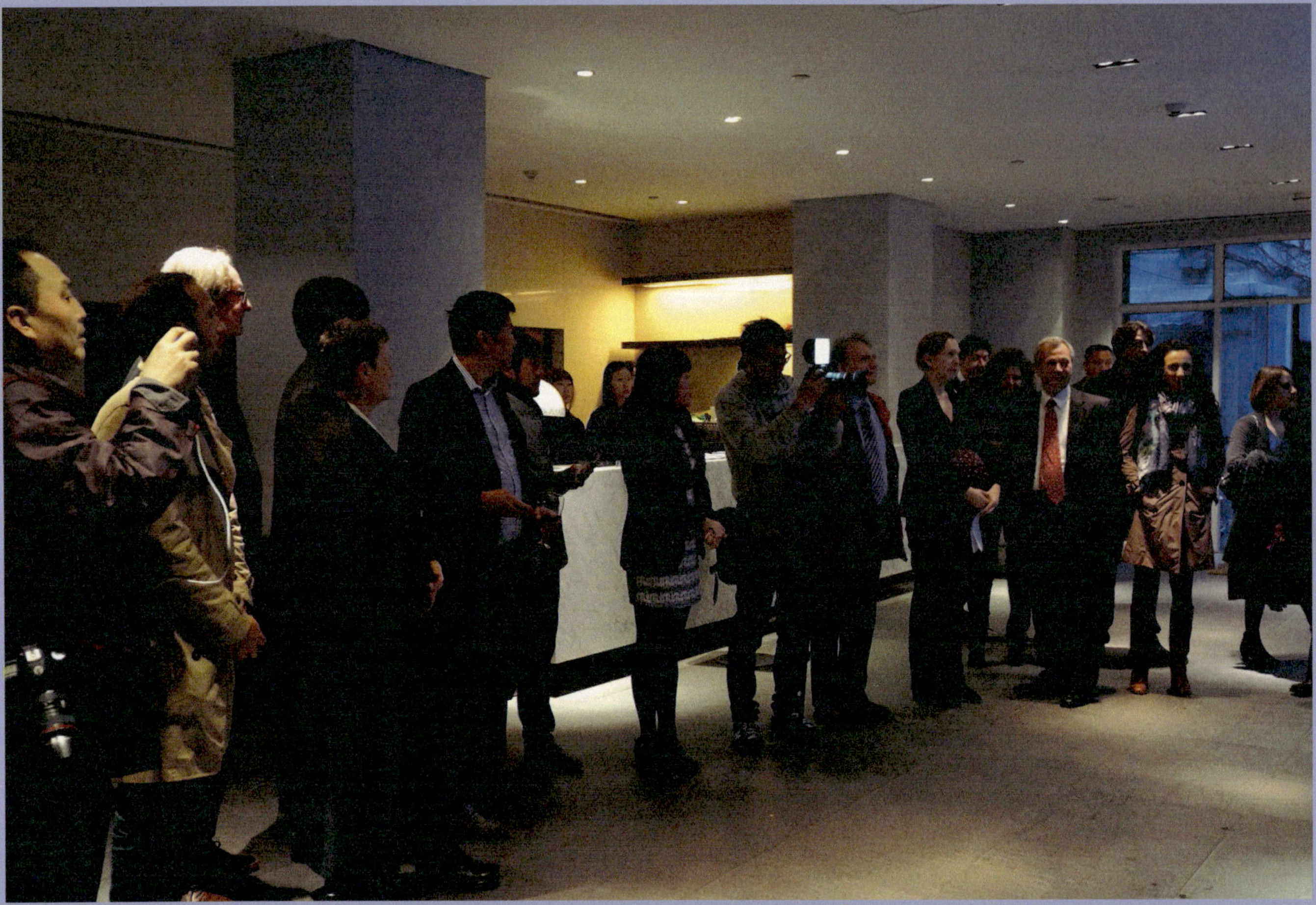

Above, LANDSEASKY's opening at OCT–Contemporary Art Terminal Shanghai; below, MAAP Korea Project Coordinator Eunju Kim.

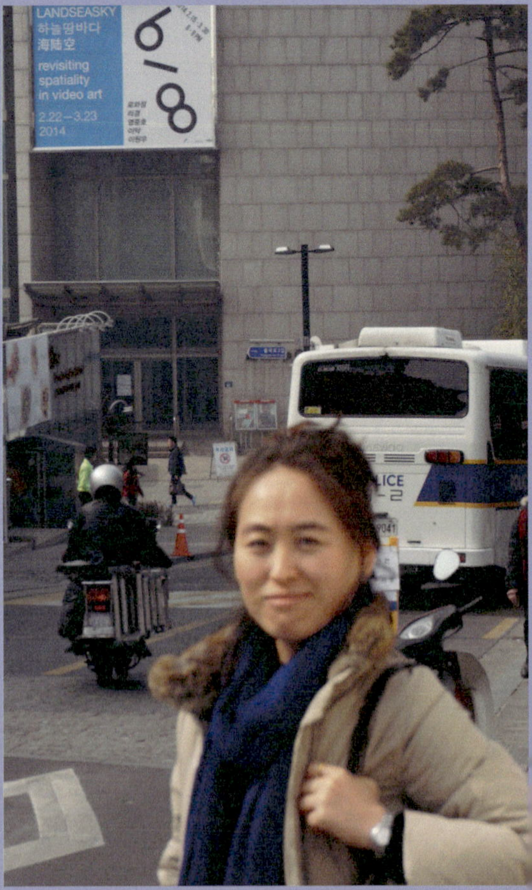

Background

MAAP is a platform for contemporary art development, presentation, and critique. Our reputation has been built on a strong portfolio of international multimedia festivals, exhibitions, and symposiums in Australia and the Asia Pacific region, locating Australian contemporary media arts within a plural international dialogue.

MAAP is recognised for critical exhibition and research initiatives based in Australia and throughout the Asia Pacific region. MAAP operates internationally as a team of agents, researchers, artists, and curators.

In addition, MAAP supports a community of visual artists with Media Bank equipment loans and online resources, which includes artist and curatorial networks.

A Brisbane based not-for-profit organisation, MAAP is dedicated to linking practitioners, curators, and organisations to resources, opportunities, and international audiences.

MAAP Board

Zane Trow - Chairperson

Dave Allen - Treasurer

Paul O'Kane – Secretary

Naomi Evans - member

Christopher Meakin – member

Jeffery Sams - member

Kim Machan - staff representative, member

MAAP Staff

Kim Machan - Director

Jessica Row - Gallery Assistant and Media Bank Officer

Eunju Kim - Korea Project Coordinator

Ashlee Sang - Gallery Assistant (QUT Internship)

Davina Li - Research Assistant

Troy Skews - Exhibition Installer

Sandra Edmondstone - Bookkeeper

Volunteers

Bridgette Liu, Matthew Allan, Catrina Udy, Sanjeev Sharma

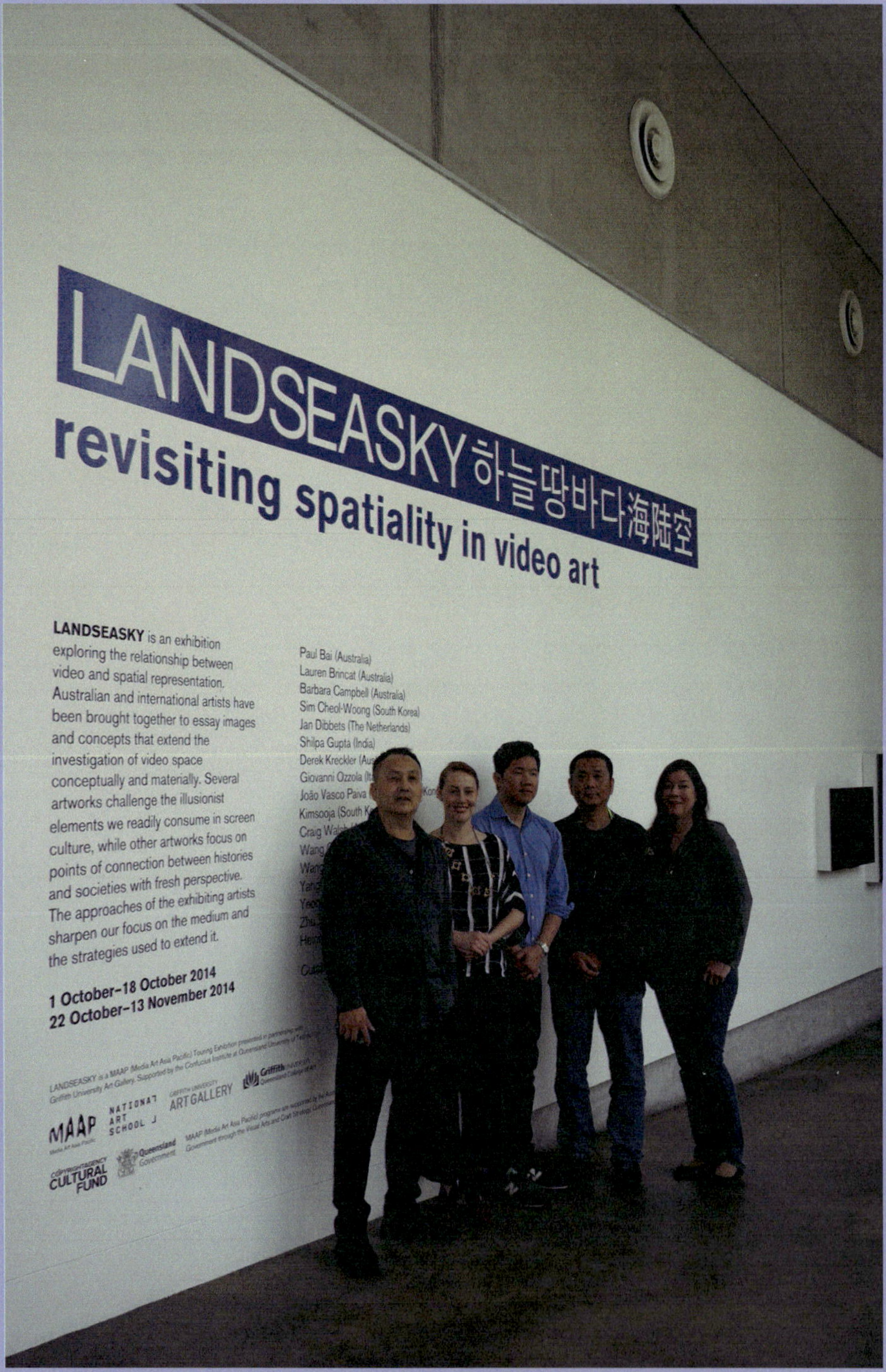

From left to right, Wang Gongxin, Naomi Evans, Paul Bai, Wang Peng and Kim Machan.

Executive Summary

After nearly two years of development the 2014 MAAP curated exhibition program was successfully managed with *LANDSEASKY: revisiting spatiality in video art (LANDSEASKY)* delivered to prestigious international museums and galleries in Korea and China before returning to Australia to be presented in Sydney and Brisbane. Commencing in February, Artsonje Center and five commercial galleries hosted LANDSEASKY in Seoul, Korea. The exhibition then travelled to China, opening in April at OCT - Contemporary Art Terminal Shanghai. In June MAAP contributed to another major international exhibition *Thingworld: International Media Art Triennial* in Beijing in partnership with The National Art Museum of China.

This is MAAP's third year of operating MAAP SPACE in Brisbane and a compelling series of exhibitions and artists' residencies were activated. Three major exhibitions were presented. The first was the extended exhibition *LANDSEASKY: Dispatches* that brought documentation of MAAP's international touring exhibition back to Brisbane audiences as a window into the project in Korea and China. This accumulative exhibition grew over the three-month duration was followed by other exhibition components of LANDSEASKY by Yang Zhenzhong (China), Giovanni Ozzola (Italy) and Sim Cheol-Woong (Korea). MAAP SPACE also hosted four artists' residencies during 2014.

MAAP generated considerable research material. In 2014, as part of the LANDSEASKY touring project, MAAP generated, organised, and participated in artists' talks, curator talks, and symposiums; these events took place in Seoul, Shanghai, Sydney, and Brisbane.

Throughout 2014 MAAP continued commitment to Sector Development through mentorships, internships, volunteer programs and very visibly through MAAP MEDIA BANK. In 2014, MAAP Media Bank supported 41 projects to a total value of $328,733. Of these, 34 exhibitions totalling $150,494 were based in Brisbane.

To date, MAAP Media Bank has supported 320 projects locally, nationally, and overseas to the equivalent commercial value now exceeding $1,500,000.

Throughout 2014, MAAP explored new models of exhibiton making through new and established partnerships while adjusting to signifcant changes to government supports for the arts. MAAP's activities in Australia, Korea and China demonstrated dynamic artistic exchange of a consistent high standard.

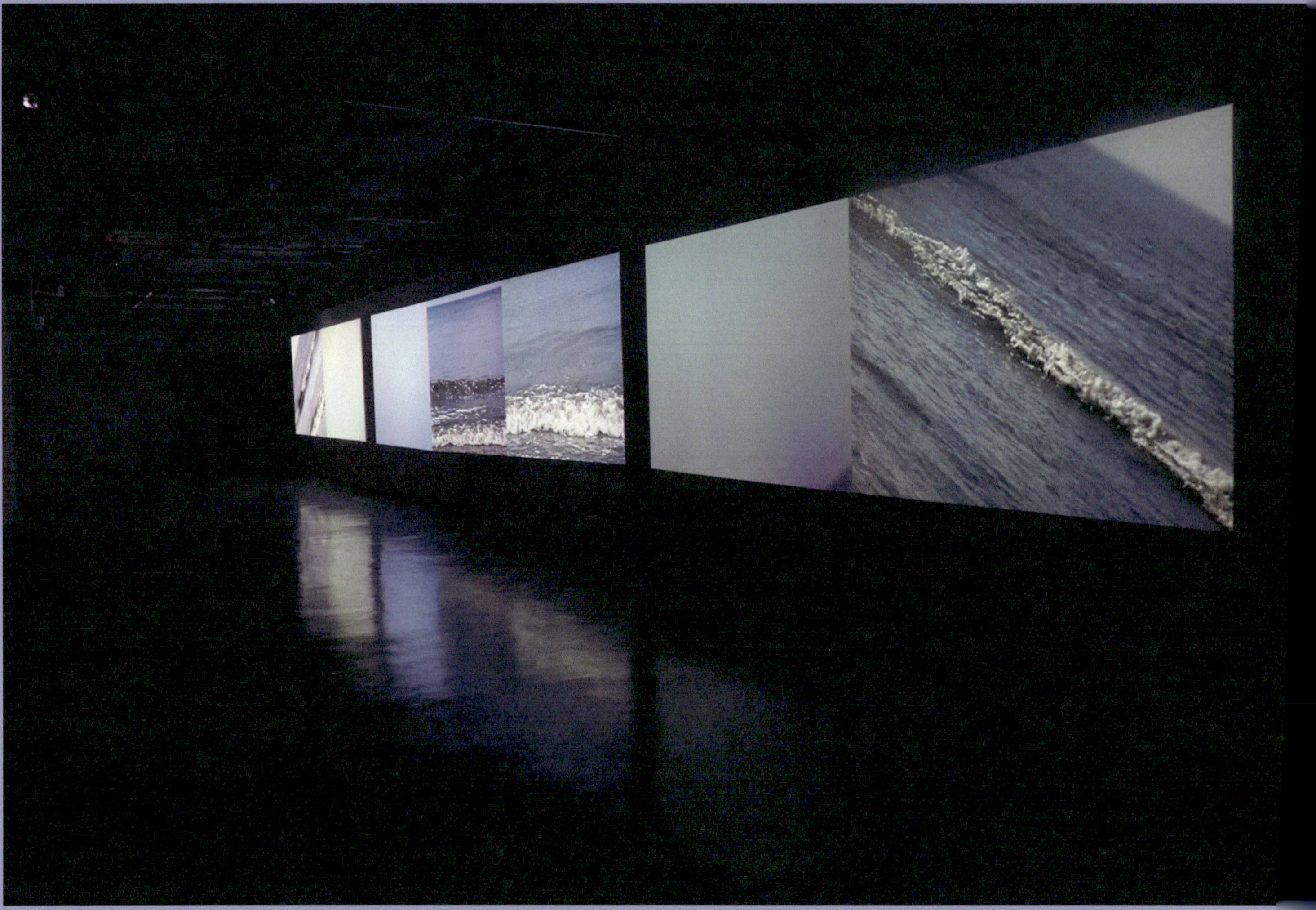

Installation view of Jan Dibbets' *Horizon I – Sea, II Sea, III Sea* at OCT–Contemporary Art Terminal Shanghai.

Content Overview:

Section 1. Core Business provides an overview of human resources, financial and accounting systems and governance (including an overview of Board Meetings held in 2014).

Section 2. MAAP Curatorial Programs Events & Projects details activities relating to the 2014 artistic program. 2014 was a highly active year with national and international exhibitions featuring the presentation of *LANDSEASKY: revisiting spatiality in video art* in three countries across ten venues. Thingworld: International Triennial of Media Art is another major 2014 international presentation realised in partnership with the National Art Museum of China. The MAAP SPACE projects discussed included satellite programs of LANDSEASKY and new work residencies that included Paul Bai, Louise Bennett, Wang Gongxin, and Wang Peng.

Section 3. MAAP Research provides information on MAAP's 2014 'Critical Dialogues' program outlined in a variety of delivery forms and venues associated with the LANDSEASKY touring project. MAAP generated, organised, and participated in artists' talks, curator talks, and symposiums; these events took place in Seoul, Shanghai, Sydney, and Brisbane.

Section 4. Sector Development Program outlines activities that have contributed to the development of artists and arts workers locally, nationally and internationally. A full report on the MAAP Media Bank supported projects is documented and valued comprehensively. A commentary on MAAP Mentorships, Internships and volunteers programs is also provided.

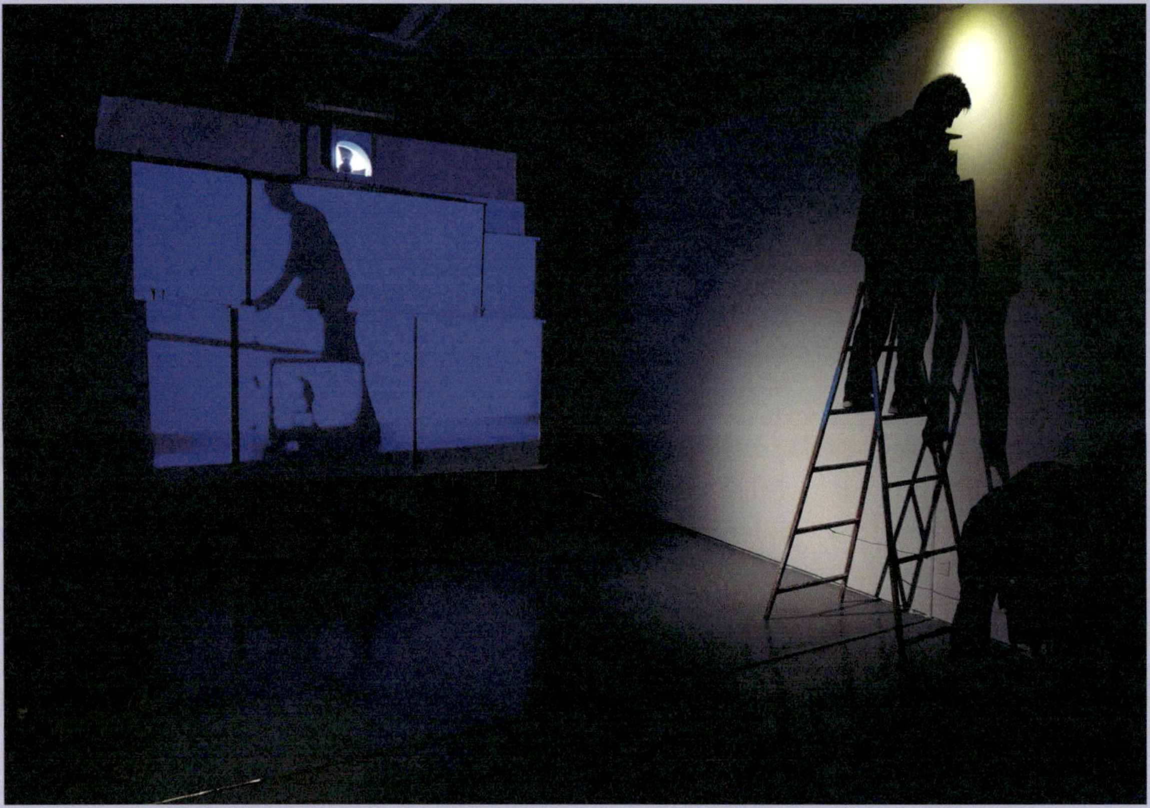

Above: workers at OCT–Contemporary Art Terminal Shanghai install Heimo Zobernig's work *Untitled Nr.23*;
below: Installation view of Giovanni Ozzola's *Garage - sometimes you can see much more.*

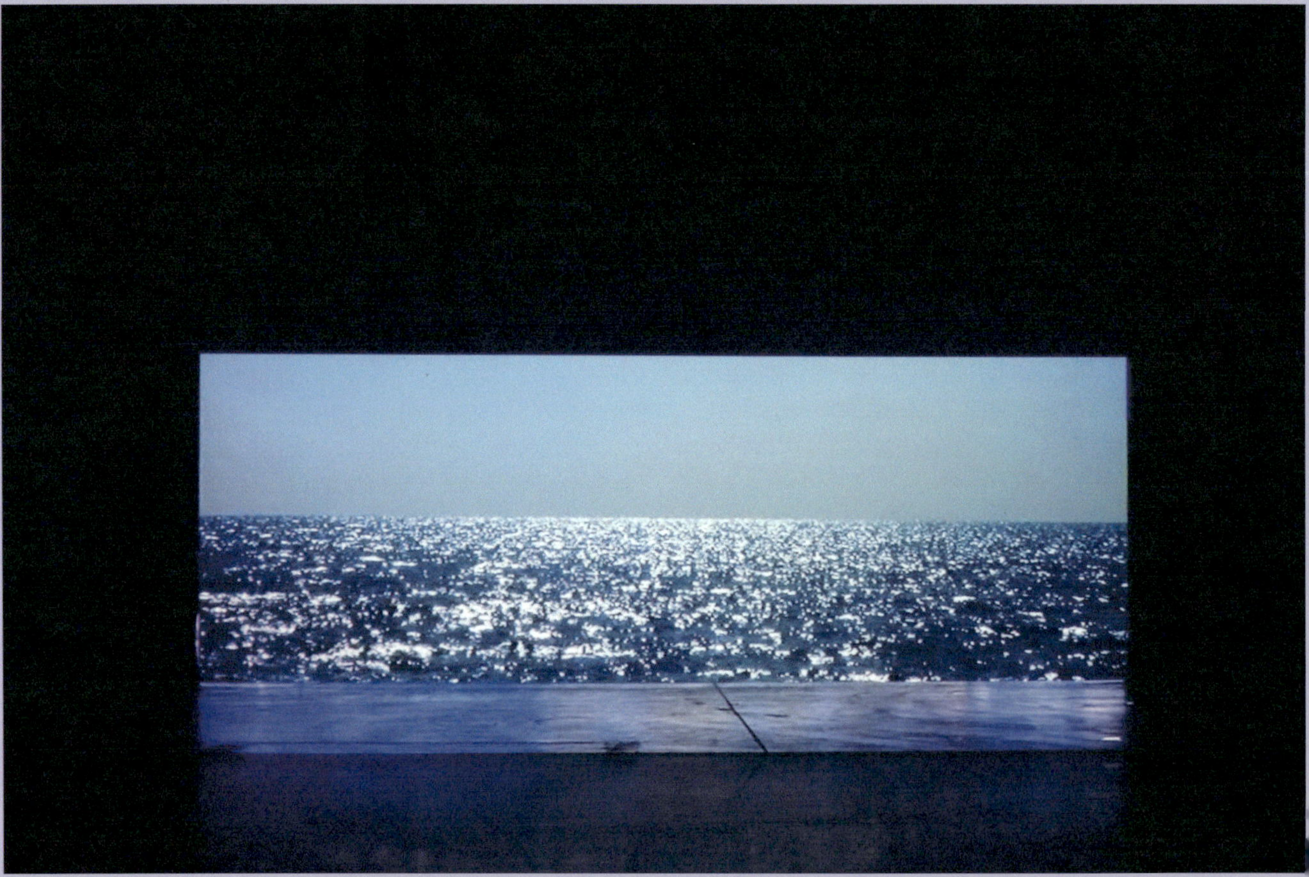

2014 PROGRAM

1. MAAP Core Business

2. MAAP Curatorial Programs, Events & Projects

- 2a.Exhibition Project: LANDSEASKY
- 2b. Exhibition Project: Thing World
- 2c. MAAP SPACE
- 2d Artists Residencies

3. MAAP Research

4. Sector Development Program

- 4a. MAAP Media Bank Report
- 4b. MAAP Consultancy: Queensland Emerging Artists, Mentorships and Volunteers

Installation view of Zhu Jia's *Beyond My Control.*

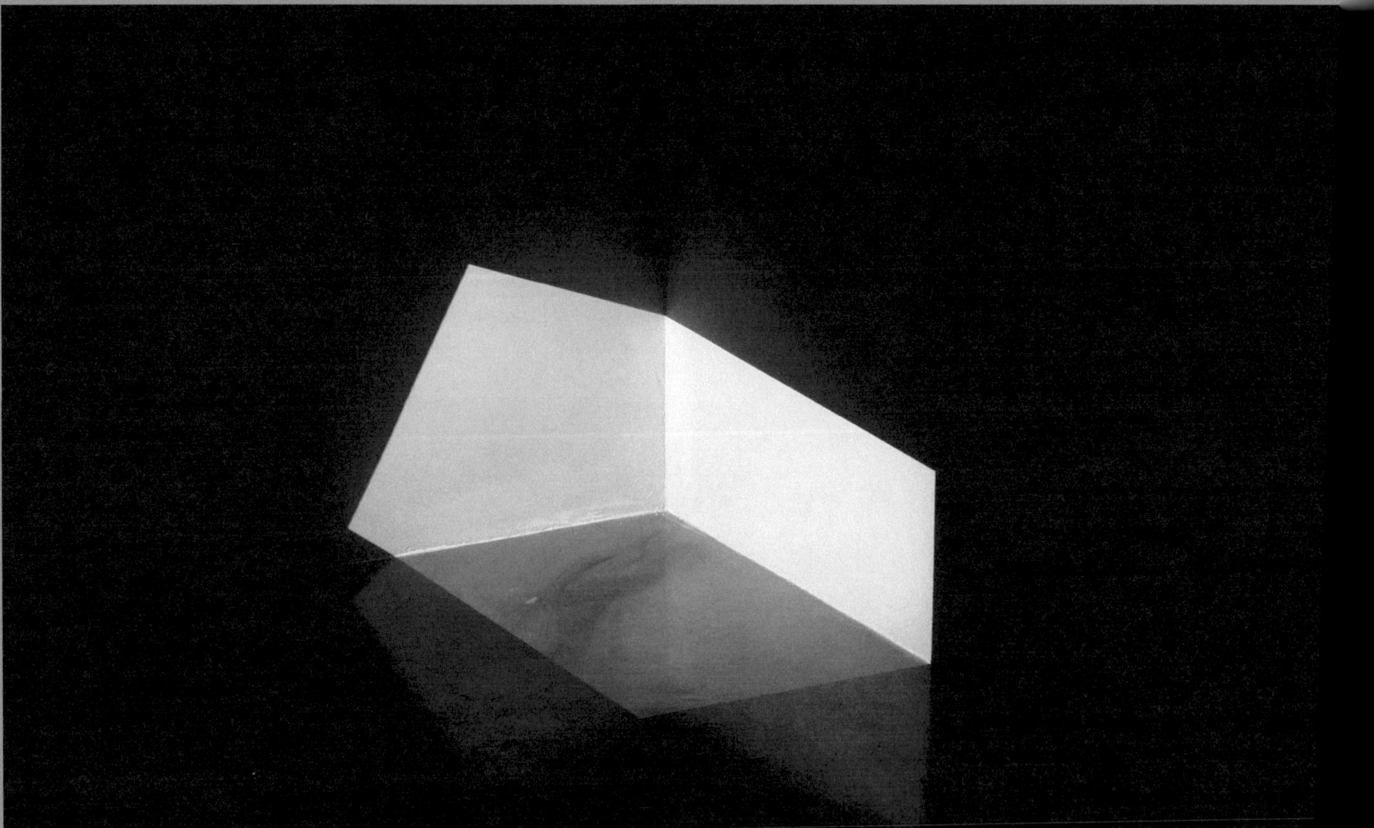

1. MAAP CORE BUSINESS

Daily operations were carried out from MAAP's office, gallery and resources centre in Fortitude Valley, led by founding Director, Kim Machan. Jessica Row continued into her second year as Gallery Assistant and Media Bank Officer. Eunju Kim commenced an internship position in August 2013 and worked on the development of the LANDSEASKY exhibition, providing Korean translation and project support to the conclusion of the Korean leg of the project in April 2014. Davina Li began as a volunteer in 2014 supporting the installation in Shanghai and also MAAP's artists' participating in 'Thingworld: International Triennial of New Media Art' at the National Art Museum of China. Four volunteers took up short-term placements over 2014 as part of the Bridgeworks program, offering a range of specialist skills including marketing, book-keeping, I.T. and design. MAAP also worked with a team of contractors throughout the year, including graphic designer Paul Bai.

Financial and Accounting System

Sandra Edmondson took over the position of MAAP's part-time bookkeeper in November 2013 and continued throughout 2014. Accounts are managed with Brian Tucker Accounting. Budgets and financial forecasting are overseen and reported to the board by MAAP Director Kim Machan and MAAP treasurer, David Allen. MAAP operates quarterly GST reporting.

Governance

Over 2014, the MAAP Board was kept in close contact with program developments, staffing arrangements, budgets and other pressing issues via regular meetings and email 'flying minutes'. In the year there were 4 meetings that met or exceeded quorum (4 members). The Board's ongoing evaluation of MAAP operations through 'monitor-review-report' strategy and 'appreciative inquiry' applied at Board meetings, ensures improved business operations. In 2014, Naomi Evans, Acting Director of Griffith University Art Gallery joined the MAAP Board.

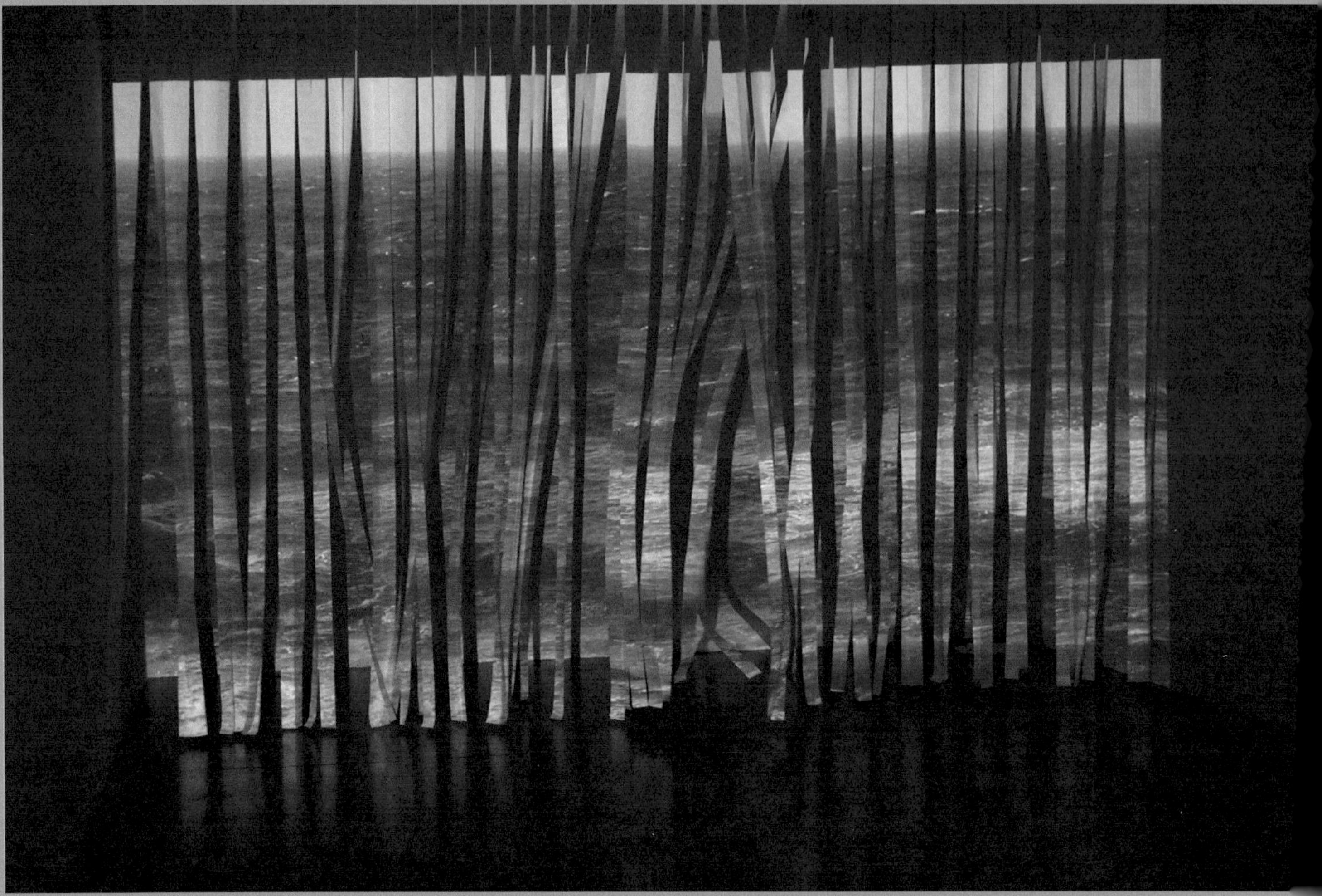

Installation view of Derek Kreckler's *Littoral*,

2. MAAP CURATORIAL PROGRAMS, EVENTS & PROJECTS

2a. Exhibition Project: *LANDSEASKY: revisiting spatiality in video art*

PROJECT OVERVIEW

2014 MAAP delivered the most extensive exhibition tour in the history of the organisation. MAAP presented and promoted LANDSEASKY, an exhibition of video and screen-media installations by 5 Australian and 12 international artists. LANDSEASKY is an international touring exhibition of video art that explores the relationship between video and spatial representation.

The selected artists works have a strong physical and visual impact and present images and concepts that extend the investigation of video space. Some challenge the illusionist elements that we readily consume in popular screen culture, while others update the connection between history and society with fresh perspective. Their approaches sharpen our focus on the video medium and the strategies that artists use to extend it.

The exhibition was launched in South Korea (22 February to 23 March 2014) across five venues in Seoul: Artsonje Center, Lee Hwaik Gallery, One and J Gallery, Opsis Art, Gallery IHN, and Gallery Skape. After Seoul, LANDSEASKY toured in China at the OCT Contemporary Art Terminal Shanghai (19 April to 22 June 2014). Following Shanghai, LANDSEASKY toured Sydney (20 August to 11 October), at the National Art School Gallery, and finally Brisbane (1 October to 13 November), at the Griffith University Art Gallery. In March 2015, LANDSEASKY will continue its international tour to China in the Guandong Museum of Art.

17 artists are exhibited in LANDSEASKY: Paul Bai (Australia), Lauren Brincat (Australia), Jan Dibbets (The Netherlands), Barbara Campbell (Australia), Wang Gongxin (China), Shilpa Gupta (India), Yeondoo Jung (South Korea), Derek Kreckler (Australia), Giovanni Ozzola (Italy), Jao Vasco Paiva (Portugal/Hong Kong), Wang Peng (China), Kimsooja (South Korea), Craig Walsh (Australia), Sim Cheol-Woong (South Korea), Yang Zhenzhong (China), Zhu Jia (China), Heimo Zobernig (Austria) Zhang Peili (China).

LANDSEASKY TOUR

Research and project development began well before the 'Light from Light' tour finished in 2012. A successful application to the Australia Korea Foundation (a section of the Department of Foreign Affairs and Trade) allowed active research that commenced with a research tour to South Korea, followed by a Curatorial Residency for MAAP Director Kim Machan in April 2012 at the Geumencheon Art Space in Seoul. Patnerships in Seoul and Shanghai enabled the venue support that was crucial to the foundation of the 2014 tour. All partners in the tour contributed invaluable knowledge and resources to the project.

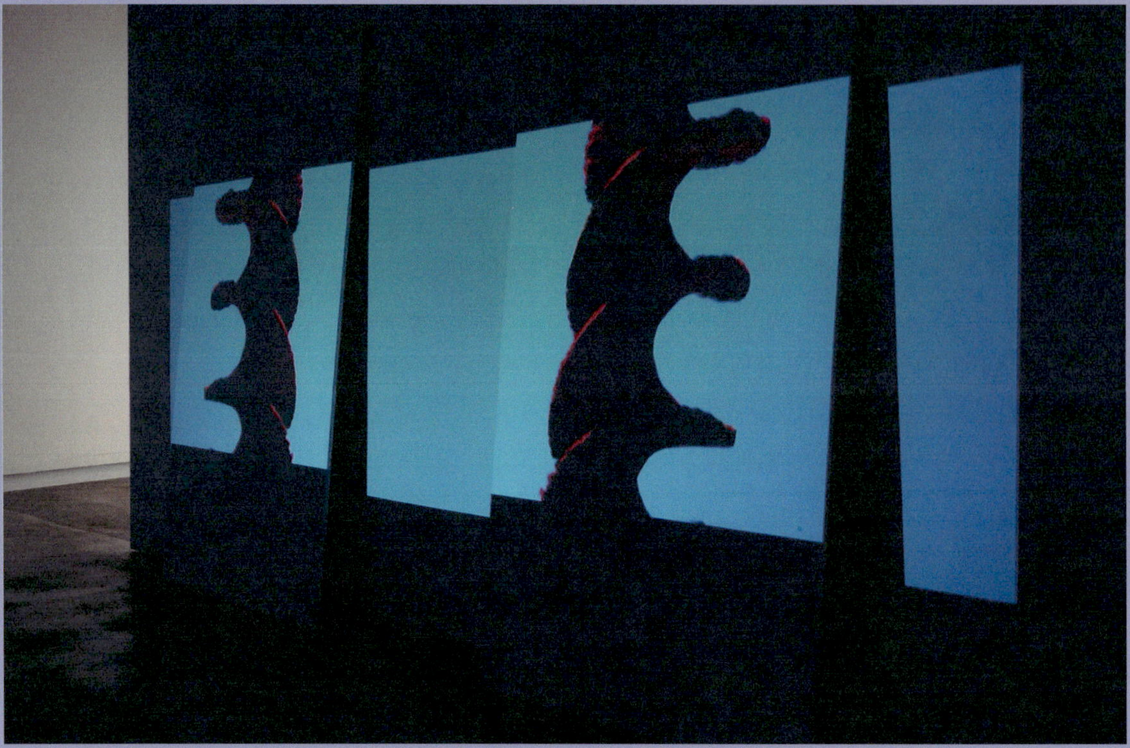

Above, installation view of Paul Bai's *Untitled (windcharm)*; below, curator Kim Machan explains artworks to the Korean media.

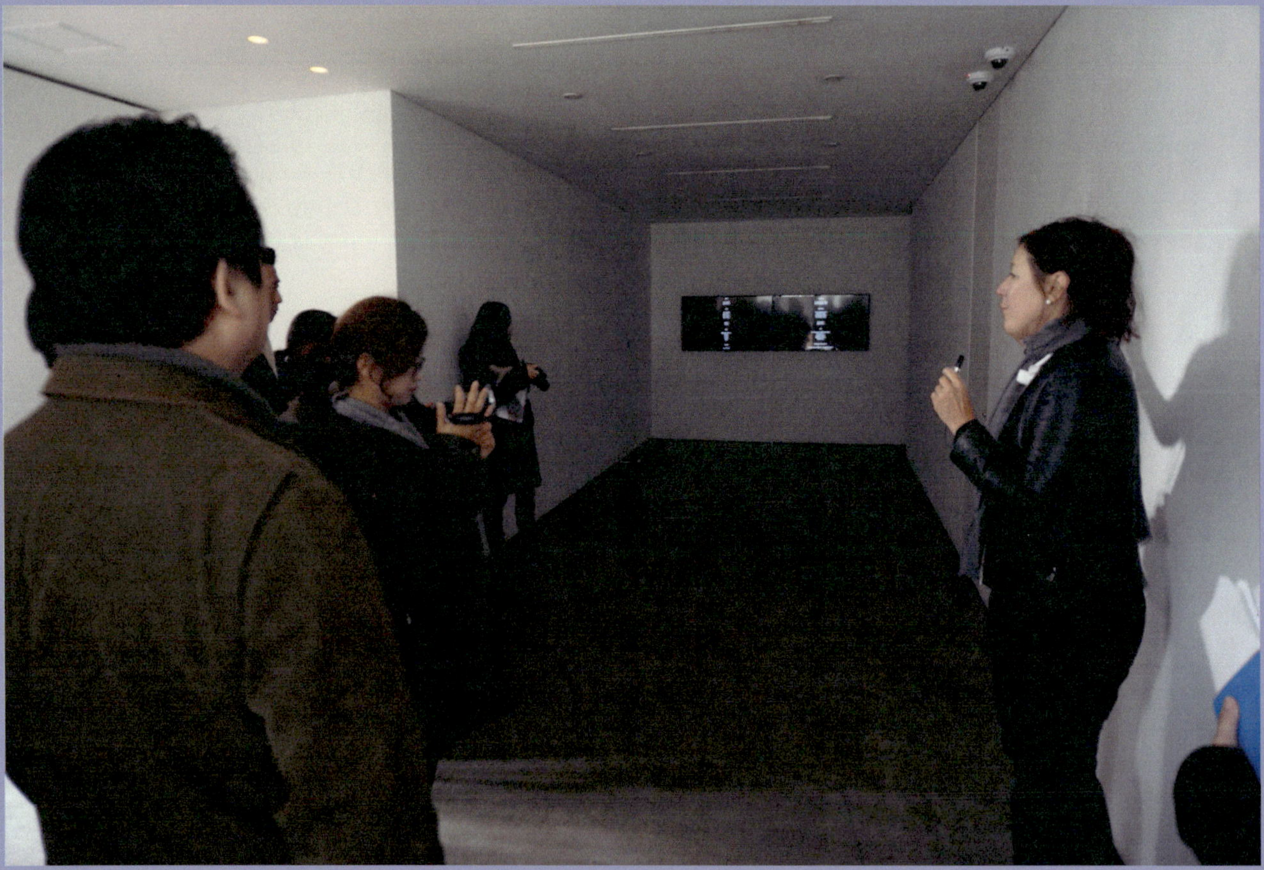

SEOUL

22 February – 23 March 2014

South Korea, Seoul venues:

- Art Sonje Center - Jan Dibbets, João Vasco Paiva, Kimsooja
- Lee Hwaik Gallery – Lauren Brincat, Yeondoo Jung, Shilpa Gupta
- ONE & J GALLERY Barbara Campbell, Derek Kreckler, Wang Peng
- Opsis Art – Craig Walsh, Giovanni Ozzola, Sim Cheol-Woong
- Gallery IHN – Paul Bai, Wang Gongxin, Heimo Zobernig
- Gallery Skape – evening window projections by Lauren Brincat and Craig Walsh

Respected Korean curator and Director of SAMUSO, Kim Sunjung, collaborated with MAAP to integrate 6 venues, five commercial galleries and Art Sonje Centre - a private contemporary art museum. The venues were located at the epicentre of the Samcheong-dong arts district and within five minutes walking distance of each other. With just three artists at each venue, works were curated to be experienced in six distinct chapters that explored the concept of spatiality in specific subsets. The walk between the spaces was very much an intended pause to consider what has been seen and anticipate the next destination. Many visitors commented that this was a very positive experience and highlighted the process of viewing provoked their understanding and thoughts about spatiality.

More than 50 media reports were documented including four major Korean television broadcasts that included general news reporting on the exhibition as well as using the exhibition as a set for a local Korean television drama. Each gallery ran its own press campaign, so the effect of 6 promotion campaigns running simultaneously created quite a buzz. The work was critically lauded in the press for the content and careful selection of art works. Social media sharing and personal blog posts have not been captured in this count.

This was the first venue of the LANDSEASKY tour and so newly commissioned works were 'world premieres' at this occasion. Included new works by: Wang Gongxin, Wang Peng, Paul Bai, Derek Kreckler, Barbara Campbell. This was also the first time that all 3 series of Jan Dibbetts 'Horizon-Sea' 1971 had ever been exhibited. Brendan Berne, Minister and Deputy Head of Mission, Australian Embassy Seoul, Republic of Korea spoke at the opening. Lauren Brincat, Barbara Campbell, Paul Bai, Craig Walsh, Derek Kreckler with assistant Jenny Tubby attended the installation and opening in Seoul along with Korean artists Yeondoo Jung and Sim Cheol-Woong.

SHANGHAI

20 April – 22 June 2014

OCT Contemporary Art Terminal Shanghai

No.1016 North Suzhou Rd. Zha Bei District

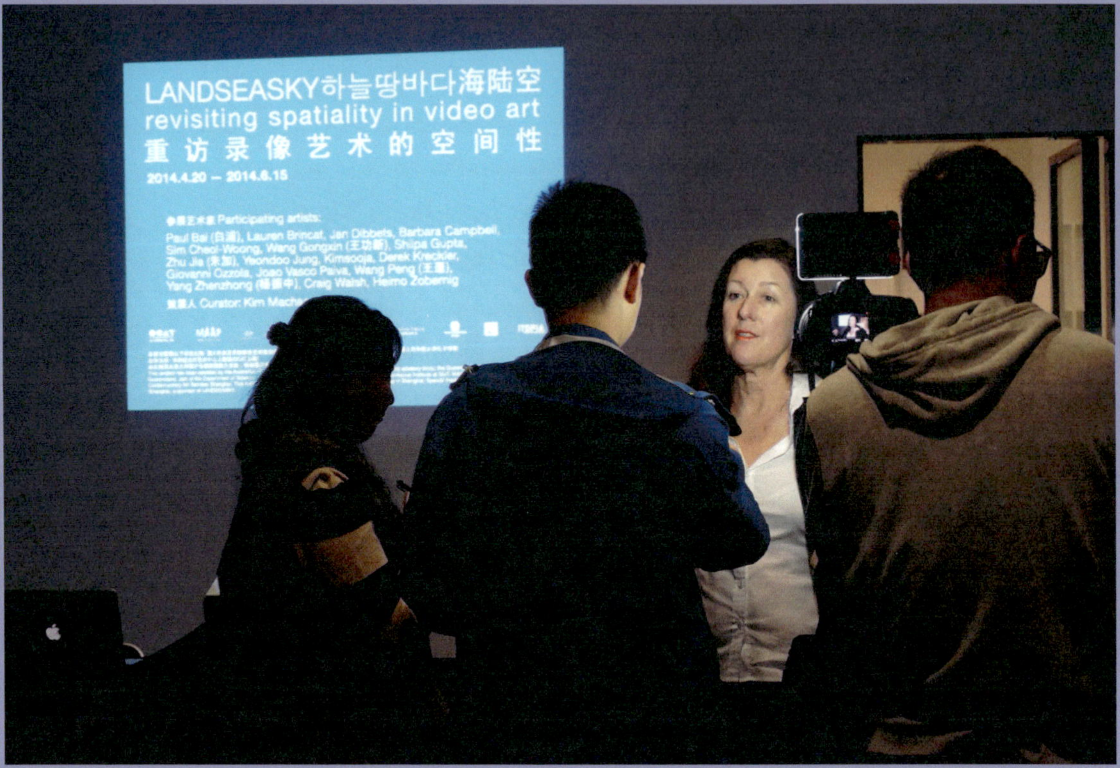

Above, Kim Machan accepts an interview from Shanghai TV; below, the opening of LANDSEASKY at OCT– Contemporary art Terminal Shanghai.

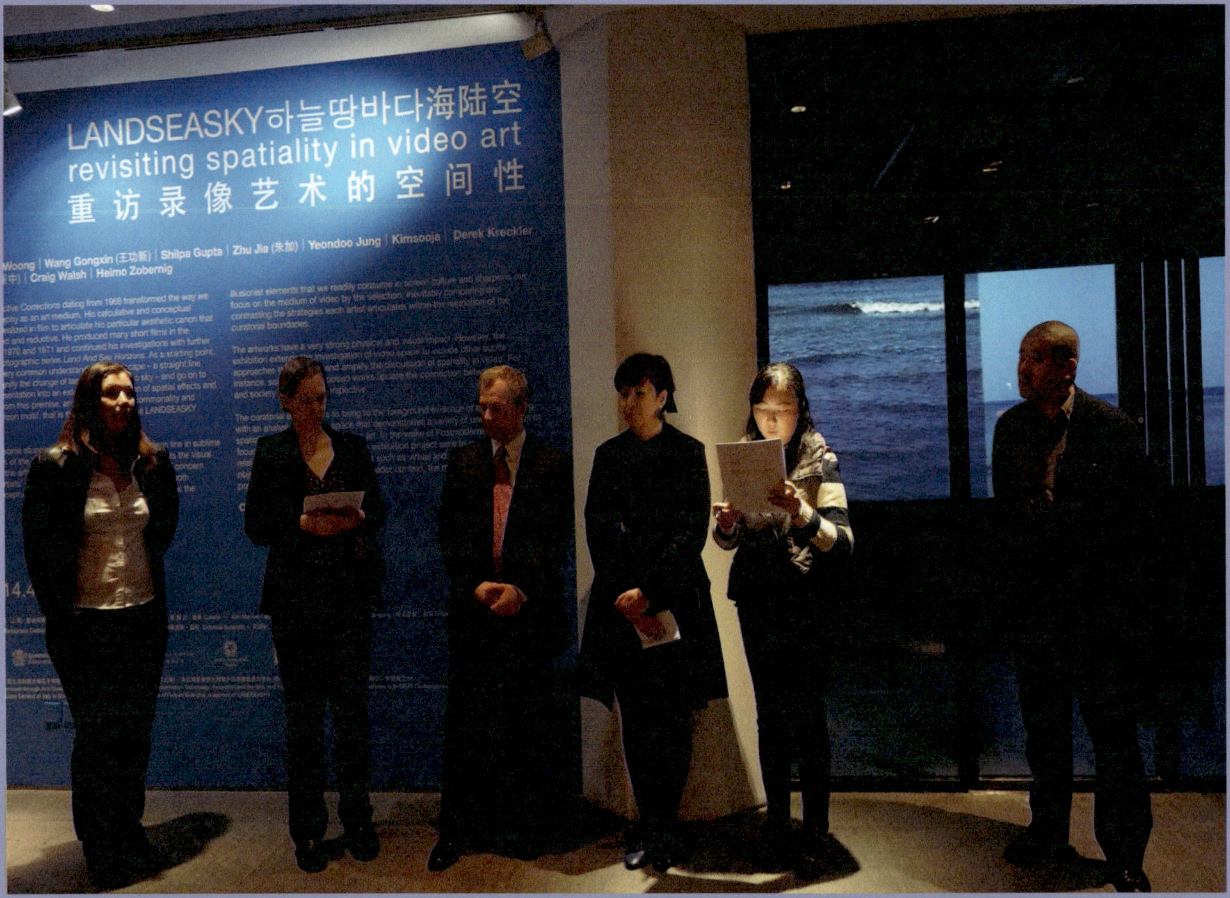

In Shanghai the exhibition was supported and presented with OCT- Contemporary Art Terminal Shanghai (OCAT) where curator, Kim Machan, was invited to design the exhibition to her own specifications and was provided with all AV equipment for the exhibits. OCAT held a press conference for 21 journalists and 2 Television crews; opening dinner for 100 guests at OCT Executive Club. Megan Jones, Deputy Consul-General Shanghai spoke at the opening. Barbara Campbell, Paul Bai, Derek Kreckler with assistant Jenny Tubby attended the installation and opening in Shanghai. Also attending were Giovanni Ozzola, Wang Gongxin, Zhu Jia and Zhang Peili. The Australian artists participated in a symposium with QUT Professor Andrew McNamara, Kim Machan and Giovanni Ozzola at The Jongi University; School of Visual Arts hosted by Hu Jieming – artist and Head of School.

OCAT strongly supported the installation and their technical support offered was extensive. The exhibition space was built to our specifications and two new artists joined the exhibition. Zhu Jia produced a new work specifically for the exhibition and Yang Zhenzong's work 'Passage' was also added. In Shanghai more than 30 media reports were documented including two major Chinese television networks, Art Forum (China Edition) and Art Forum International. Because of the location of the OCAT, the audience in Shanghai was more 'art world' concentrated and made up of artists, critics, academics, and students.

SYDNEY

20 August – 11 October 2014

National Art School Gallery

Forbes Street, Darlinghurst, Sydney

There were three private previews with National Art School supporters, sponsors and curators from the Tate Modern. The opening hosted an enthusiastic crowd of about 200 and was officially opened by Liz Anne McGregor Director MCA Sydney. Paul Bai, Derek Kreckler, Barbara Campbell, Lauren Brincat attended the opening. The art works were installed in the 19th Century sandstone building originally built by convicts as a goal and now home to the National Art School. This new partnership worked well and the exhibition garnered great praise. The exposure of MAAP through the exhibition in Sydney was a positive lift to awareness of our most recent international projects. National Art School Director Judith Blackall said the exhibition was well attended attracting above average audience numbers and that people were spending longer than average time viewing the exhibition.

The National Art School gallery building has very strong architectural elements and the installation design took advantage of some of the extreme possibilities of the space. This included placing Zhu Jia's work at the top of a very dramatic staircase. The upper floor was particularly popular with reports from the gallery that people returned to enjoy the installation and also to have their lunch

There was a particularly strong volunteer support group connected to the gallery and their contribution to the installation details was really admirable and enhanced the experience for all. Artist Mike Parr commented in the attendance book "splendid exhibition! The top floor is particularly marvellous. Dibbets, Kreckler, magical!".

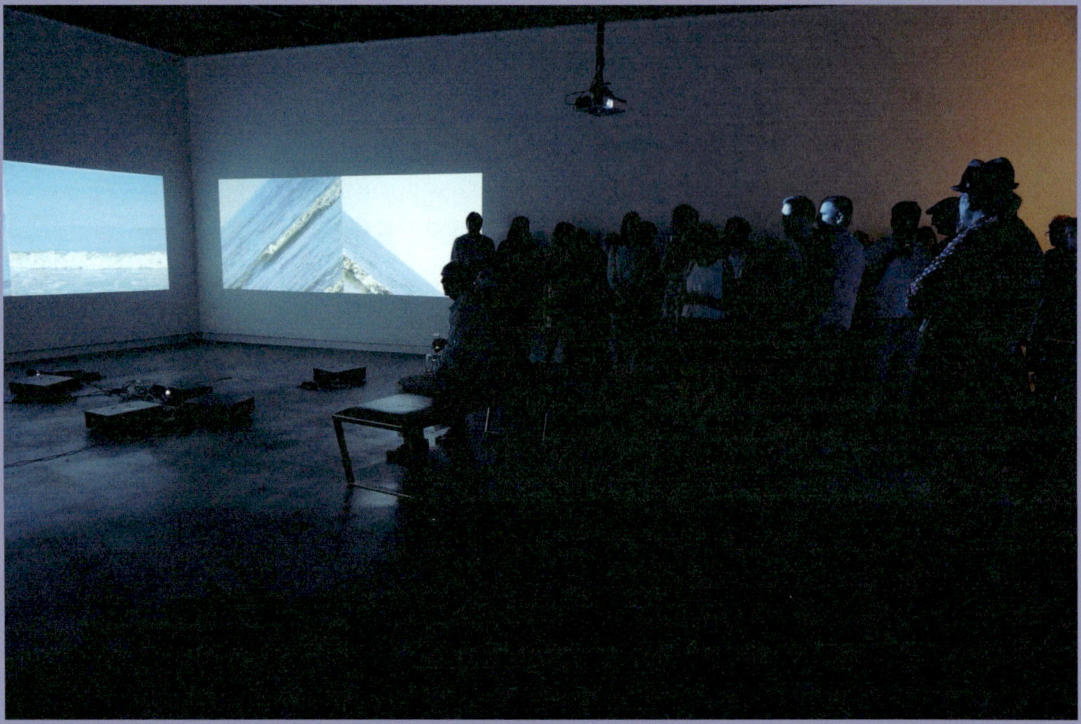

Above and below, LANDSEASKY opening at Griffith University Art Gallery.

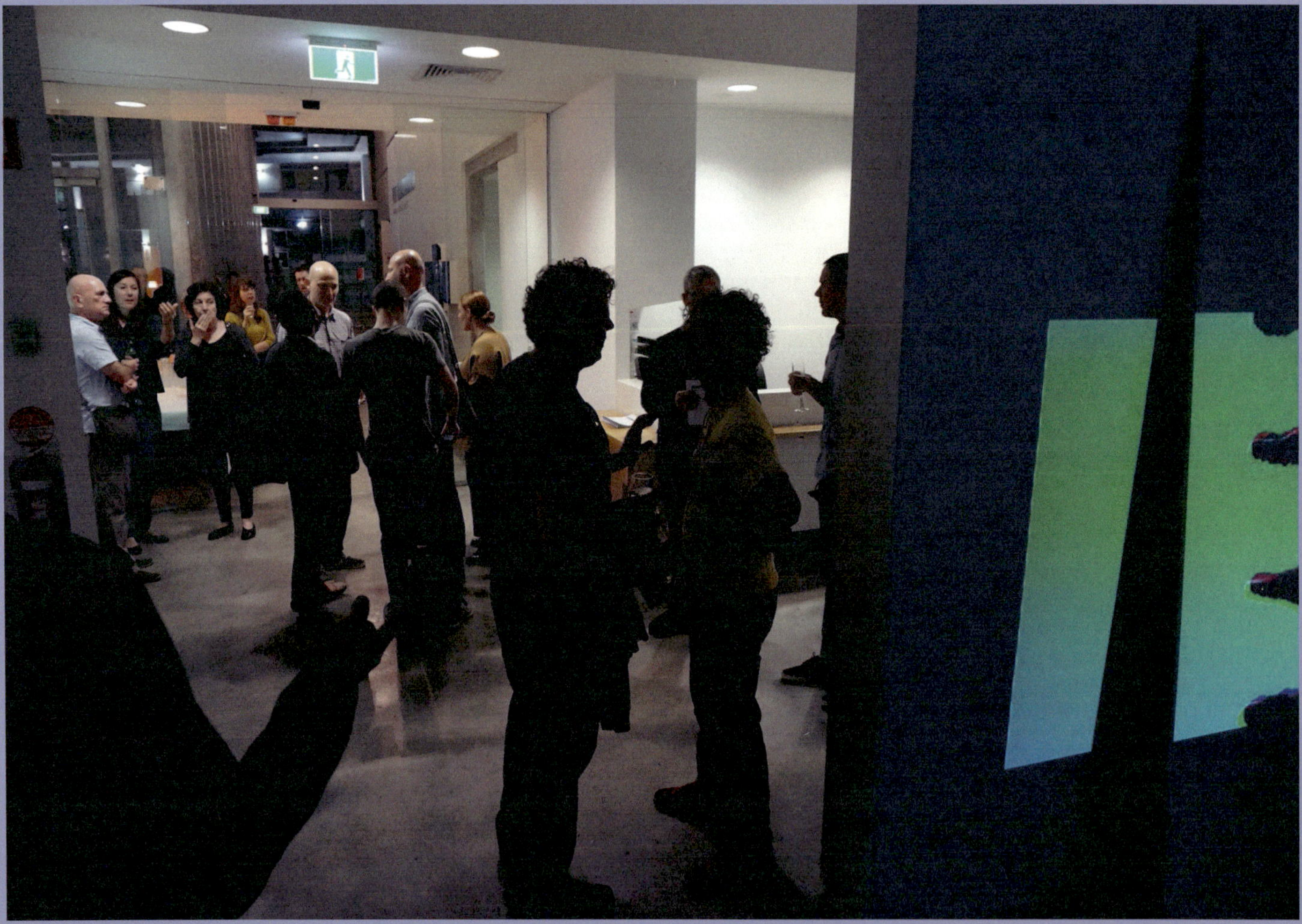

BRISBANE

Chapter 1: 1 October – 18 October 2014
Chapter 2: 22 October – 13 November 2014
Griffith University Art Gallery
Queensland College of Art, South Bank, Brisbane

2 parts + 2 venues; having some works remain in the venue Griffith University Art Gallery, as a founding partner in the LANDSEASKY exhibition applications and tour supporter, enthusiastically supported the project and endorsed the art school gallery partnership with the National Art School Gallery Sydney. There was collaboration with marketing, promotion and design, and was the first project that directly connected these two galleries.

Due to the specific nature of each supporting venue, new approaches were taken with each installation design and configuration. In this instance a quite different exhibition design was required and works were shuffled in two time frames and two venues. The exhibition was not simply divided there were artists that straddled both configurations. Jan Dibbets', Wang Peng and Zhu Jia's work stayed in place across both Chapters in the Griffith University Art Gallery while other works in the first Chapter were by Heimo Zobernig, Wang Gongxin and Paul Bai. These artists were replaced by Derek Kreckler, Barbara Campbell and João Vasco Paiva.

Russell Storer, Curatorial Manager of Asian Art QAGoMA (Queensland Art Gallery and Gallery of Modern Art) officially opened the exhibition. The second chapter Artists present for the installation and opening were Paul Bai, Derek Kreckler, Barbara Campbell, and from China, Wang Gongxin and Wang Peng. The Chinese artists expanded their activities in Brisbane to take up 'artist in residence' status. Wang Gongxin had a very specific project that he wanted to achieve while here in Brisbane and Wang Peng had a more open intentions (see detail in section 2d 'Artists Residencies'). MAAP SPACE hosted additional artists yang Zhenzhong, Giovanni Ozzola and Sim Cheol-Woong. (see page 27)

FUTURE ACTIVITIES

Site visits to Guangzhou have occurred and negotiations continue. MAAP is working with Guangdong Museum of Art to present *LANDSEASKY: revisitng spatiality in video art* in their museum in 2015 across an entire floor that exceeds 2,000 square metres. MAAP has commenced a trilingual catalogue that will document all venues and hold scholarly essays; commentary and photographs of the works in situ.

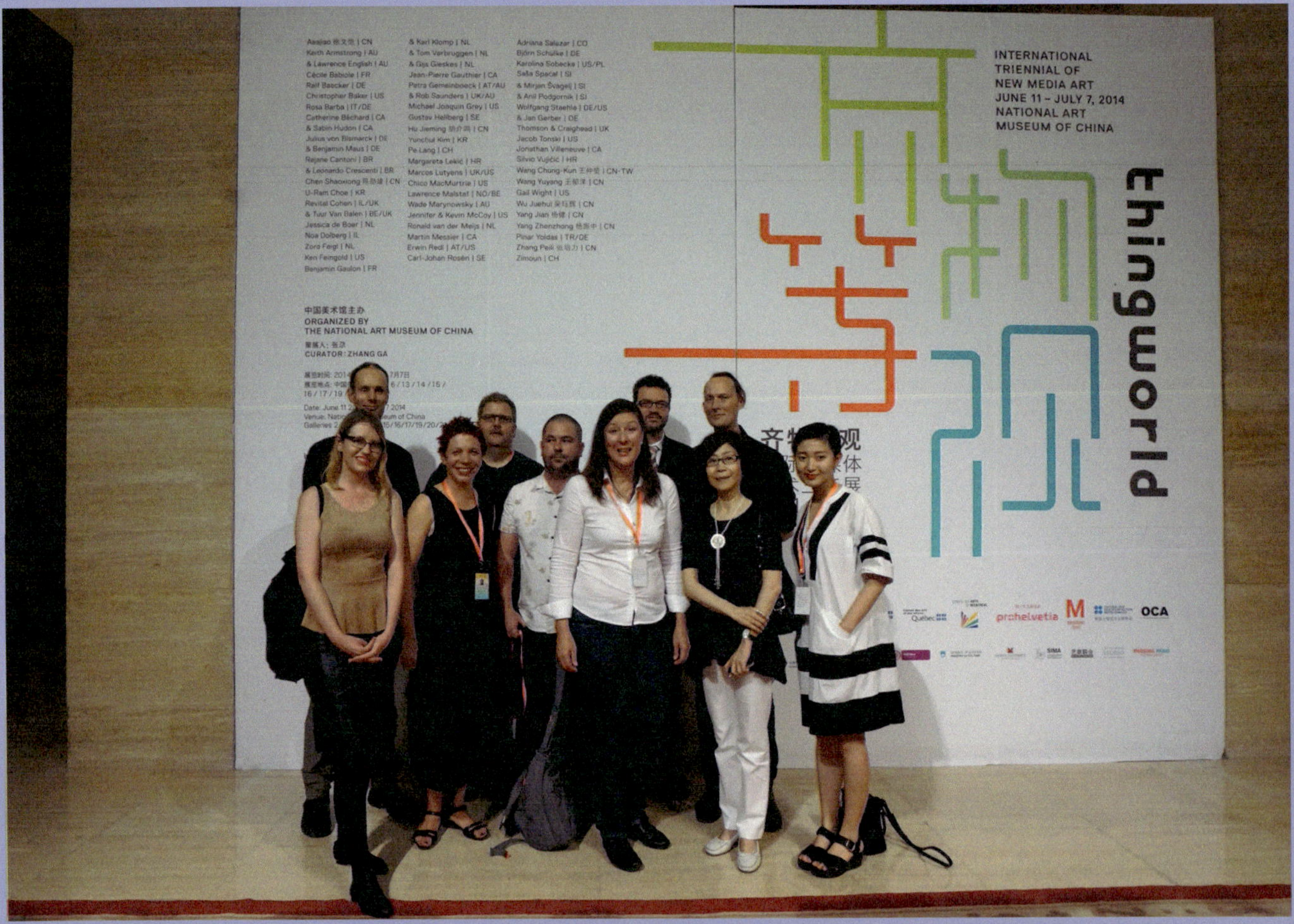

Curator Kim Machan, Davina Li and Australian artists Wade Marynowsky, Keith Armstrong, Lawrence English, Petra Gemeinboeck, and Rob Saunders with staff from the Australian Embassy Beijing.

2b. Exhibition Project: Thingworld: International Triennial of Media Art

10 June – 7 July 2014
The National Art Museum of China,
1 Wusi Daijie, East District, Beijing, China, 100010

MAAP was invited to partner with the National Art Museum of China (NAMOC) again in 2014*. Australian artists Keith Armstrong, Laurence English, Petra Gemeinboeck, Rob Saunders, and Wade Marynowski presented three major art installations. Zhang Ga, the Artistic Director for *Thingworld*, chose to focus on how media art can rethink subjectivity and equality by examining the autonomy of nonhuman agency. Working with the exhibition's theoretical theme, MAAP supported three works. Wade Marynowsky's "Acconci Robot" 2012 is an interactive robot that follows people unawares. Keith Armstrong and Lawrence English's "Light of Extinction" 2012 is an elaborately constructed low light and immersive audio environment, created from an autonomous gaggle of robotic levers and pulleys. Petra Gemeinboeck and Rob Saunders' "Accomplice" 2013 features two robotic 'creatures' that were installed behind a constructed wall in the gallery. "Accomplice" was equipped with a camera and a hammer that interacted, by playfully hitting and progressively puncturing the wall, with the audiences' movements. These three works—Light of Extinction, Accomplice, and The Acconci Robot—participated in one of the most expansive media art exhibitions of its kind: 70 artworks, by over 65 artists from 22 countries, were featured. The exhibition occupied multiple gallery spaces across 6,000 square metres and was open in Beijing from 10 June to 9 July 2014.

With funding from the Experimental Art Fund of the Australia Council, MAAP facilitated the China presentation, installation, de-installation, and return of the artists' works. MAAP organised all freight logistics, made travel arrangements for the five artists who attended, and prepared written material for the catalogue. MAAP also prepared press kit information on each artist's work and liaised with promotional staff from NAMOC.

There were 65 artists in the exhibition and over 70 major art works were presented. Next to MAAP, other international partners of this this prestigious media art exhibition included V2 Institute for Unstable Media, Rotterdam; Transmediale Art and digital culture, Berlin; FACT, Liverpool; and the Academy of Arts and Design Tsinghua University. There were 28 contributing supporters including several countries embassies that all attended the opening ceremony. The Australian Embassy attended with five members of staff including Michael Growder, Jennifer Mason, Hayley Ward Senior Cultural Relations Officer, May Lea Ling, and previous staff Amanda Barry who has since left but still holds strong ties after working with MAAP since 2008. They showed great interest in the exhibits and were helpful and enthusiastic.

Wade Marynowsky, Keith Armstrong, Lawrence English, Petra Gemeinboeck, and Rob Saunders all travelled to Beijing to supervise the installation of their works and attend the opening ceremony, press conference, and symposium. All artists were involved in press interviews and television news stories.

* MAAP previously partnered with the National Art Museum of China in 2011 with *Light from Light* exhibition and 2008 with Synthetic Times International Biennial of Media Art.

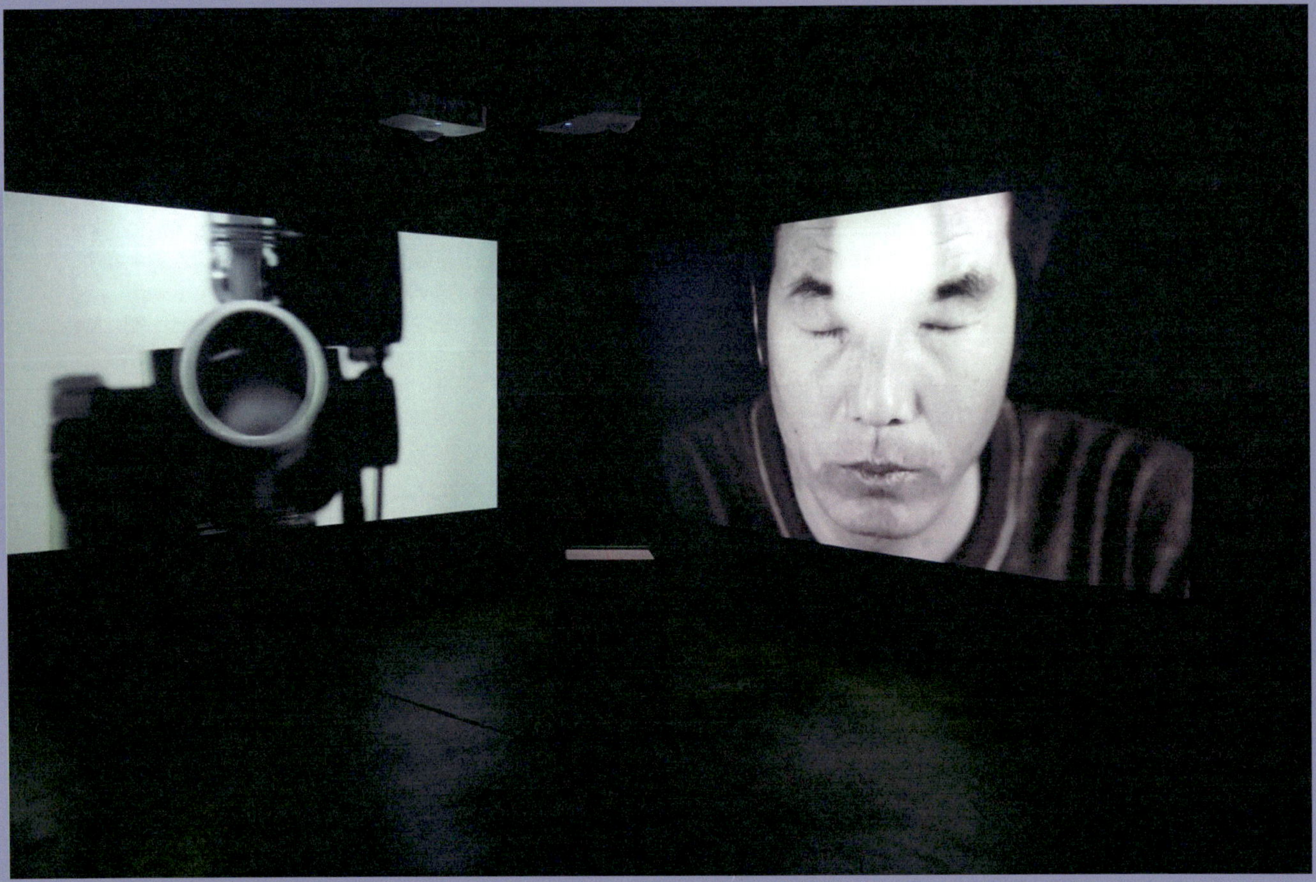

Installation view of Wang Gongxin's *The Other Rule in Ping Pong.*

2c MAAP SPACE

MAAP SPACE EXHIBITION 1

LANDSEASKY Dispatches 23 May – 30 September 2014

LANDSEASKY Dispatches exhibition was a collection of moving and still documentation and collateral from the various venues that accumulated during the tour. The gallery provided a window to the international activities of MAAP for Brisbane audiences and demonstrated the scope of the installation internationally. The exhibition of documentation was built upon during the exhibition period as new venues were occupied and additional documentation forwarded. LANDSEASKY Dispatches presented a collection of 280 images of the artworks installed in Shanghai (at OCT Contemporary Art Terminal Shanghai), Sydney (National Art School Gallery) and the collection of six galleries in Seoul including Artsonje Center, ONE AND J Gallery, Gallery IHN, Gallery SCAPE, OPSIS Art, and Lee Hwaik Gallery. The collection of material in this evolving exhibition kept local audiences informed and acted as a great tool to talk about MAAP's extended activities internationally.

Opening event and curators talk: 23 May 2014

MAAP SPACE EXHIBITION 2

1 October – 28 October 2014

Works by Yang Zhenzhong and Giovanni Ozzola were played sequentially in 20 minute sequences.

Yang Zhenzhong

As part of LANDSEASKY Chapter 1 in Brisbane, MAAP SPACE presented Yang Zhenzhong's video installation, *Passage* 2012. MAAP first exhibited artworks by Yang Zhenzhong in 2000 and provided exposure of his work to the Australian contemporary art audiences. In recent years, the Queensland Art Gallery of Modern Art (QAGoMA) acquired work by Yang; *Passages* represents one of his most recent productions and was well received. The work was installed with a robust immersive audio experience.

Public event and curators talk: 24 October 2014

Giovanni Ozzola

Giovanni Ozzola's video installation, *Garage - sometimes you can see much more* 2009 – 2011, was presented at MAAP SPACE as part of LANDSEASKY Chapter 1 in Brisbane. Following Ozzola's Australian 'debut' at the Sydney LANDSEASKY show, MAAP was proud to introduce Ozzola to Brisbane audiences. This work embodies a central practice of Ozzola: the exploration of three-dimensional space and its relationship to light. The video sequence of a roller door rising and falling dramatically alters the viewer's perception of the gallery space.

Public event and curators talk: 24 October 2014

MAAP SPACE EXHIBITION 3

Sim Cheol-Woong 29 October – 11 November 2014

The installation of An|other River, 2011, three-channel projection was an Australian premiere. The artist was first exhibited in the Gold Coast City Gallery at an Australia-Korea Cultural Exchange exhibition where MAAP supplied sponsored display equipment. This led to further research and critical engagement with the artist's practice. The installation at MAAP SPACE worked particularly well with the strong architectural characteristics of the main gallery.

Public event and curators talk: 7 November 2014

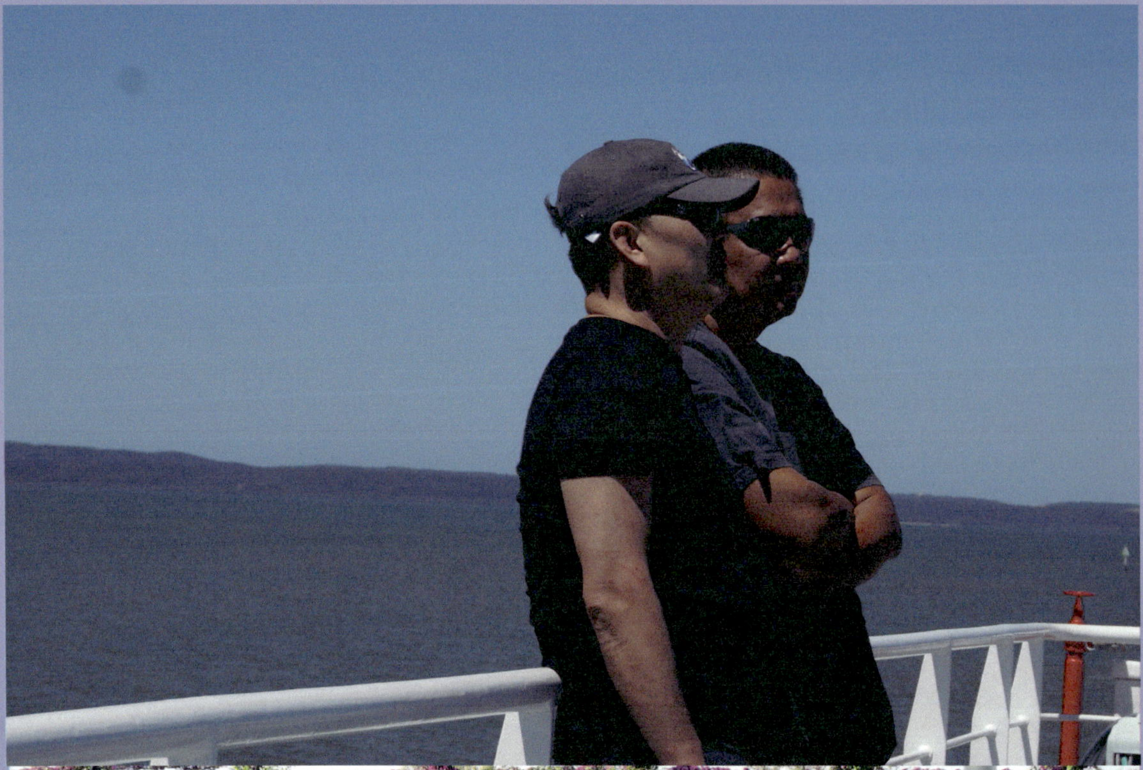

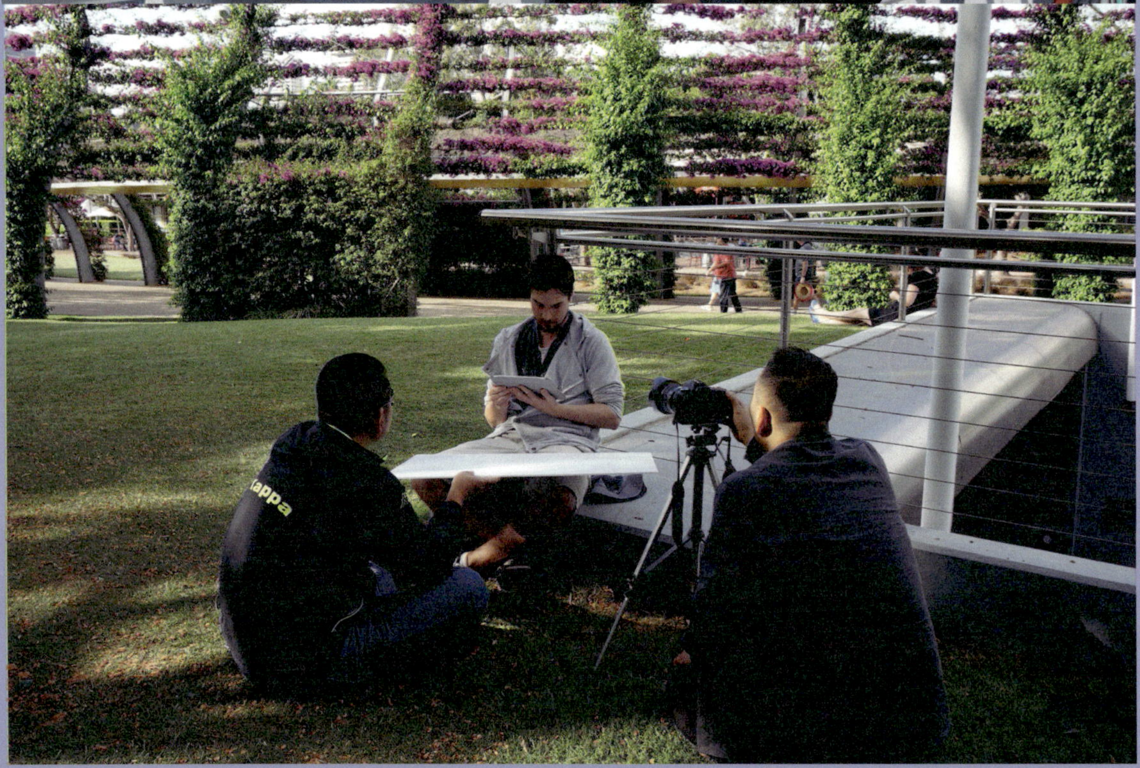

Above, artists Wang Gongxin and Wang Peng travel to Stradbroke Island; below, the artists record footage during their residency at MAAP.

2d Artist Residencies

MAAP SPACE RESIDENCY 1
Paul Bai February 2014

In preparation for the LANDSEASKY exhibition, Paul Bai used his residency to test several versions of his video installation. He used projectors and media players to develop the new work. He also used this time to explore other mediums for forthcoming projects. Bai said, "It was really very useful to be able to directly test installation configurations. Each exhibition site is different, but many issues were resolved through tests made at MAAP SPACE".

MAAP SPACE RESIDENCY 2
Louise Bennett October 2014

During October Louise Bennett came to test equipment and view work in progress.
The residency was low key and intermittent and demonstrated other models of support that MAAP can give artists. This residency reminds us that artists use a variety of spaces to develop work and that MAAP can be very responsive and useful with direct access to a space with a full complement of AV media on hand.

MAAP SPACE RESIDENCY 3
Wang Gongxin 1 – 7 October 2014

Through funding from the Queensland University of Technology's (QUT) Confucius Institute, MAAP was able to invite Chinese artist Wang Gongxin, based in New York City and Beijing, to participate in the Brisbane LANDSEASKY opening, related events, and, most importantly, produce new artwork in an Australian setting. At the opening of LANDSEASKY, academic art-historian and art critic, Dr Sally Butler interviewed Wang Gongxin for her review on LANDSEASKY for ArtLiink.

MAAP Research Assistant, Davina Li, assisted the artist and organised all aspects of the artist's filming across five locations in inner city Brisbane and Stradbroke Island.

Davina Li arranged filming permissions and organised volunteers and actors for the production that included:

The actors involved in Wang Gongxin's new project were:

Luke Weier, The University of Queensland,

Selina Vicenzino, Somerville House,

Melina Wightman, Queensland University of Technology,

Jonathan Hickey, Griffith University, and

Matthew Meintjes, The University of Queensland.

These volunteers gained experience in working with an internationally and critically acclaimed contemporary video artist.

The results of this residency are planned to be included in his solo exhibition at OCT Contemporary Art Terminal Shanghai in March 2015.

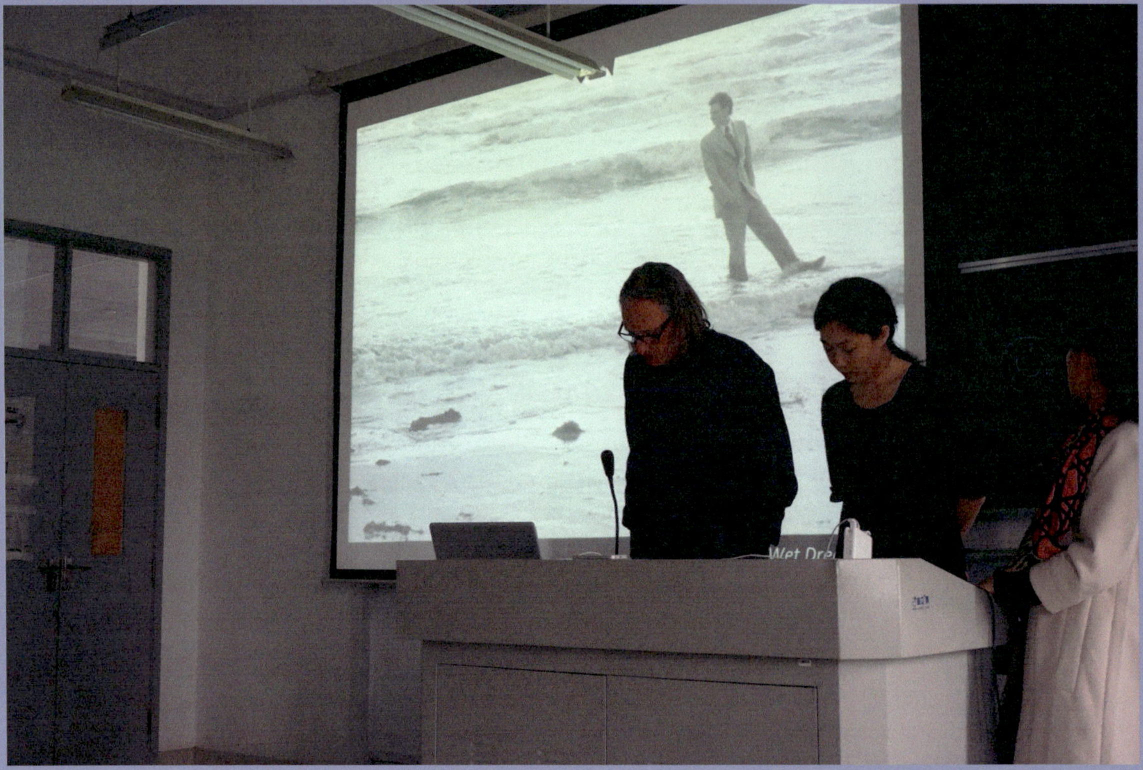

Above and below, LANDSEASKY symposium at Shanghai Institute of Visual Art.

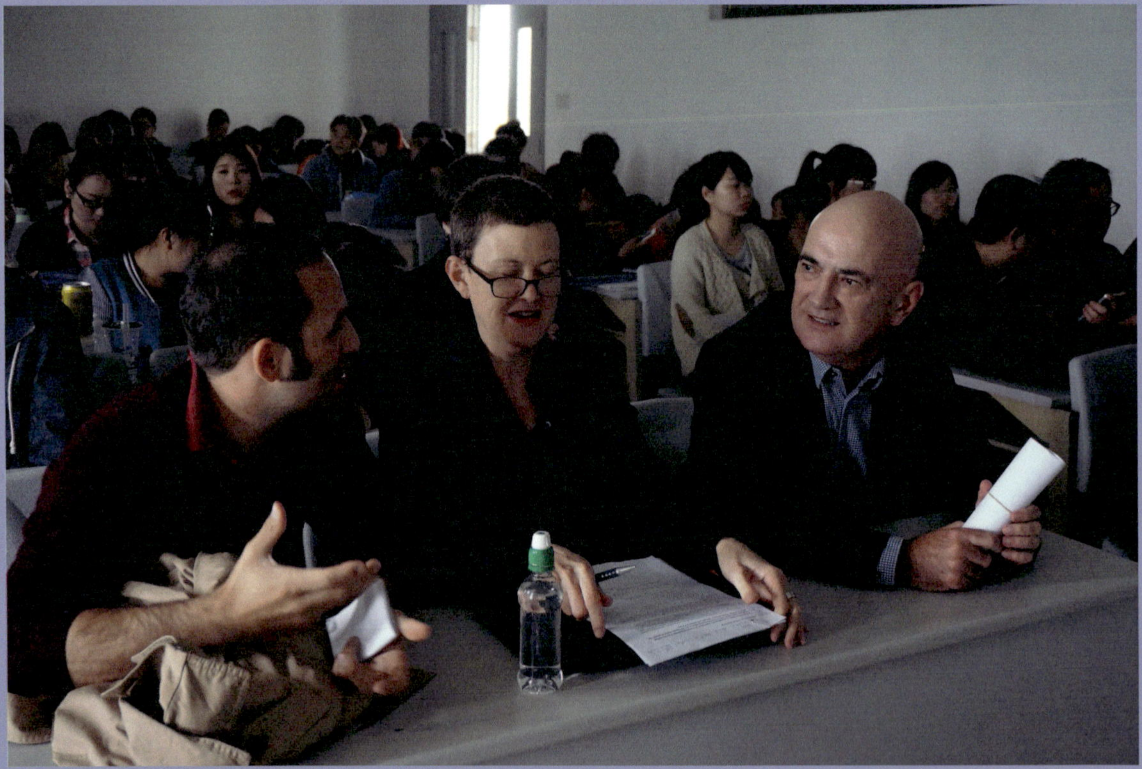

MAAP SPACE RESIDENCY 4

Wang Peng 1 October 2014 to 7 October 2014

Through funding from the Queensland University of Technology's (QUT) Confucius Institute, MAAP was able to invite Chinese artist Wang Peng, based in New York City and Beijing, to participate in the Brisbane LANDSEASKY opening and related events. This was Wang Peng's first visit to Australia and he participated in public programs for LANDSEASKY. Wang Peng met with the public, artists, academics, including Asc. Professor Rex Butler and Dr Sally Butler, and was introduced to the cultural landscape in Australia. During his residency, Wang Peng visited Stradbroke Island to film, make photographs and to develop ideas for future exhibitions.

3. MAAP RESEARCH

Critical Dialogues

In 2014, as part of the LANDSEASKY touring project, MAAP generated, organised, and participated in artists' talks, curator talks, and symposiums; these events took place in Seoul, Shanghai, Sydney, and Brisbane.

Shanghai Symposium

On 18 April 2014, MAAP partnered with OCT Contemporary Art Terminal Shanghai and The Shanghai Institute of Visual Culture, Fudan University to organise a LANDSEASKY symposium, which was moderated by Hu Jieming, Head of SIVA Academic Committee. After Kim Machan overviewed the curatorial rationale to LANDSEASKY, the following papers were presented:

"Modernism after Postmodernism in the Explanation of Contemporary Art ", by keynote speaker, Professor Andrew McNamara (Head of Discipline, Creative Industries Faculty, School of Media, Entertainment Creative Arts, Visual Arts, Queensland University of Technology);

"Defining the Third Spatial Position: A Brief Exploration of Spatial Representation in Contemporary Art Practice", by LANDSEASKY artist Paul Bai;

"Into Proximity" new spatial relations (an artistic research project Australia, Korea, China), by LANDSEASKY artist Barbara Campbell;

"The Zone Between", by LANDSEASKY artist Derek Kreckler; and

"Light and Video Space", by LANDSEASKY artist Giovanni Ozzola.

Over 100 students and staff attended this academic symposium, which was a great opportunity for MAAP to establish new cultural networks in China.

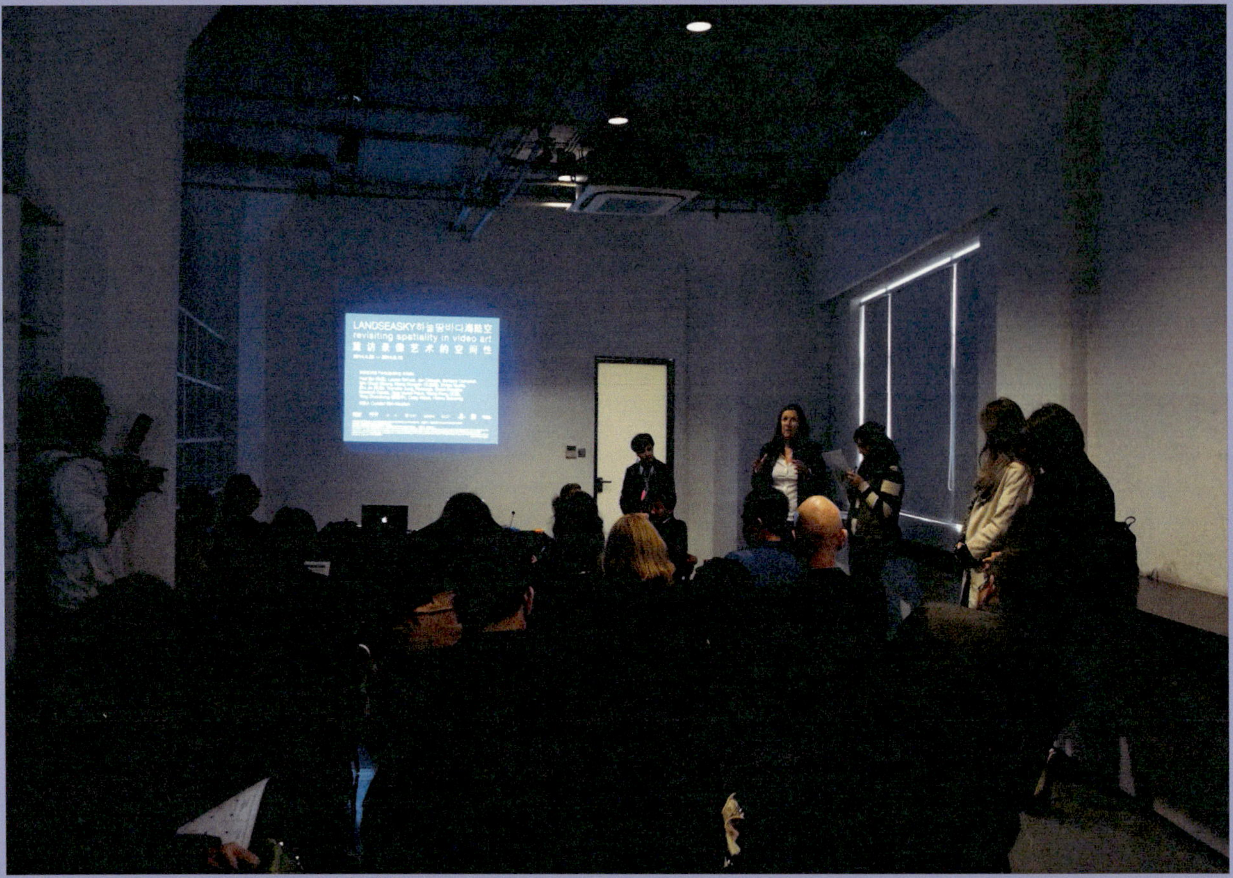

Above, LANDSEASKY media conference at OCT Contemporary Art Terminal Shanghai: below, group photo of the speakers and translators in the LANDSEASKY symposium at Shanghai Institute of Visual Art.

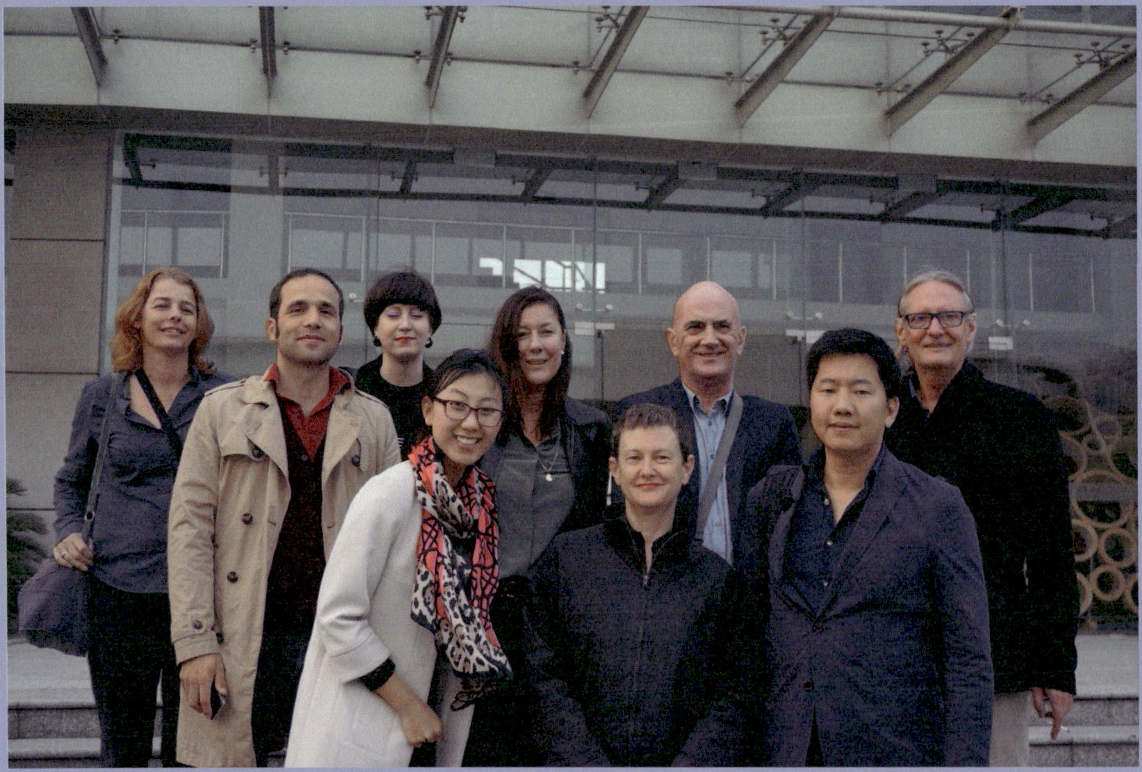

Summary of Artists' Talks

Seoul:

Attending artists were Paul Bai, Lauren Brincat, Barbara Campbell, Derek Krekler (assisted by Jenny Tubby), Sim Cheol-Woong, Craig Walsh, and Yeondoo Jung. While there were no formal artists' talks in the galleries, there was great interaction on an informal basis among artists, gallerists, curators, and attendees.

Shanghai:

The artists Paul Bai, Barbara Campbell, Derek Krekler (assisted by Jenny Tubby), Wang Gongxin, and Zhu Jia presented talks to press, students, Shanghai art academics, government officials, and guests to the opening night.

Sydney:

National Art School Sydney, 30 August, 2.30-4.00 pm

Artists attending the opening were Paul Bai, Lauren Brincat, Barbara Campbell, and Derek Kreckler (assisted by Jenny Tubby). Artists talked informally on the opening night and engaged with attendees.

Brisbane: Griffith University Art Gallery, Brisbane, 2 October, 12pm. Speakers: Wang Gongxin, Wang Peng, Paul Bai, and Kim Machan. GUAG. LSS Chapter 1

Griffith University Art Gallery, Brisbane, 23 October, 12pm. Barbara Campbell, Derek Kreckler, Kim Machan. LSS Chapter 2

Summary of Curator Talks

Seoul: Press launch and talks on the opening day and it was attended by 28 press companies and three television companies.

Shanghai: Two hour talk that was well attended by press, students, university staff, and government officials. Two television crews were involved in recording the talk and tour, which was given in English and translated into Mandarin. This event garnered positive response to the exhibition, which resulted in over 20 electronic and print articles and 3 television broadcast features.

Sydney: (1) a 'lunchbox' talk that over 80 students attended (2) a post-graduate talk (3) a talk for a law firm who sponsored the National Art School (NAS), and (4) a special preview for curators from the Tate Modern and other NAS sponsors.

Brisbane: Saturday, 11 October, 2014 at the Griffith University Art Gallery. The talk was well attended and particularly well represented by art history staff from the University of Queensland Staff, including Asc. Professor Rex Butler, Dr Sally Butler, and Dr Allison Holland. Curatorial staff from the Queensland Art Gallery/Gallery of Modern Art also attended.

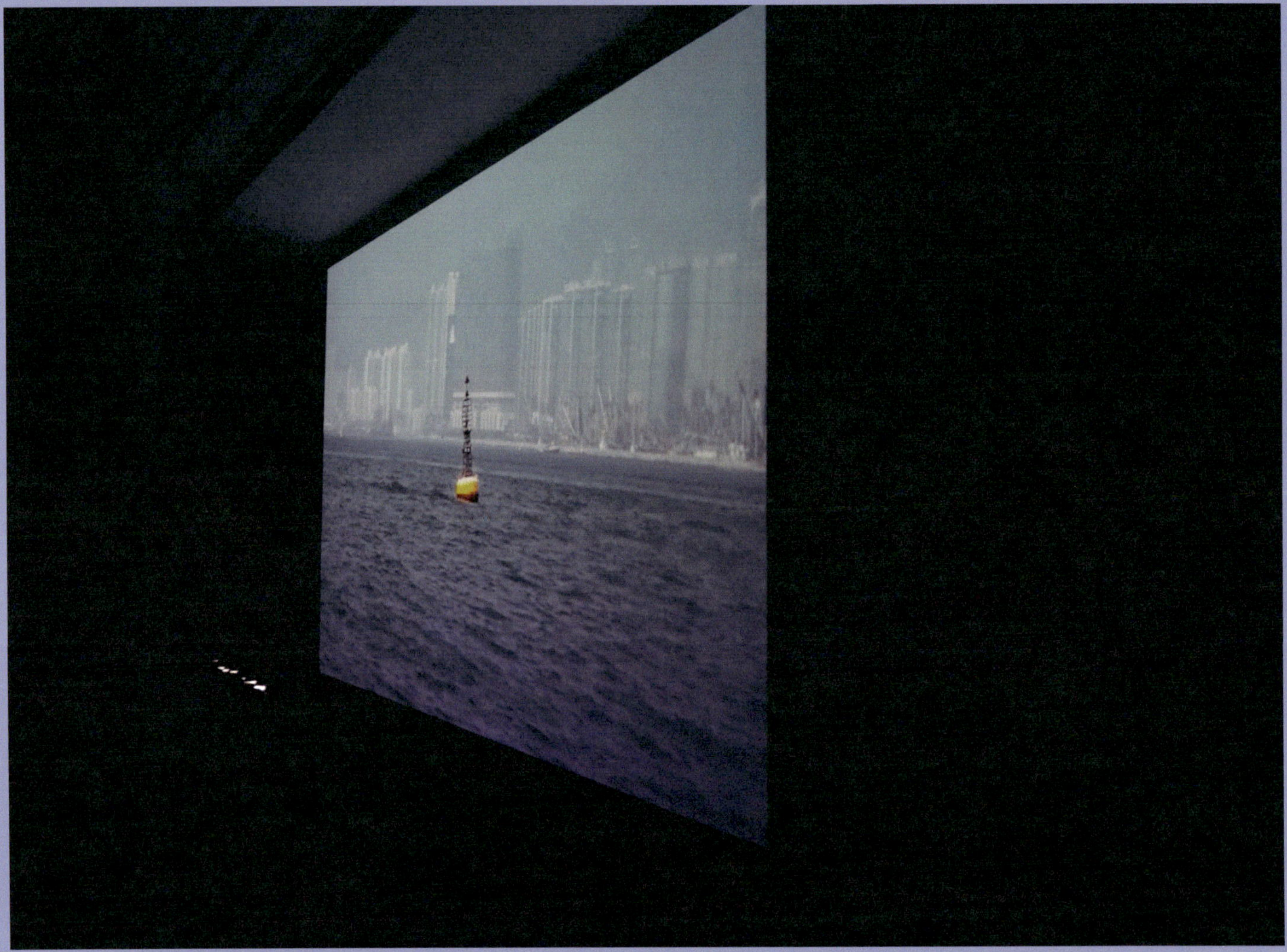

Installation view of João Vasco Paiva's *Forced Empathy*,

4 SECTOR DEVELOPMENT PROGRAM

4a. MAAP Media Bank Report

Since 2006, the MAAP Media Bank Program has offered loans of audio visual equipment, repurposed from its own exhibitions, to the local and national artistic community. To date, MAAP Media Bank has supported 320 exhibitions/projects locally, nationally, and overseas to the equivalent commercial value of approximately $1,500,000.

MAAP Media Bank is a member-based loan program designed to support local and national artists and arts workers in their creative production and events by addressing the considerable, and often prohibitive costs associated with producing and exhibiting media art.

MAAP Media Bank is the only free loan service of its kind in Australia, offering audio-visual equipment and technical consultation to the artistic community. MAAP Media Bank supports free of charge, projects of high artistic merit that align with MAAP's own curatorial endeavours, with priority given to emerging artists, artist-run-initiatives, and non-profit creative endeavours.

In 2014, MAAP Media Bank supported 41 projects to a total value of $328,733.00. Of these, 34 exhibitions totalling $150,494 were based in Brisbane.

The Gambling Community Benefit Fund grant enabled the purchase several new items of electronic equipment, which have been invaluable in the program's busy MAAP Media Bank loan schedule this year. Purchased equipments summary is outlined below.

Quantity	Description
6	Mitsubishi Electric WD3300 Data Projector
1	Mitsubishi UD740U Data Projector
3	NEC NP-M260WSG Data Projector
3	Panasonic PT-AR Data Projector
2	NEC PE401HG DLP Data Projector
7	Sony Data Projector Model VPL-CX63
1	Mitsubishi LCD Monitor 42"
3	Sony 42" Full High Definition LED LCD TV
2	NEC VT695 projector bulbs
2	Panasonic PT-AR projector bulbs
1	Apple Macbook Pro Laptop
2	Apple Mac Mini Computer
8	Brightsign HD220 Network Digital Media Player
8	Western Digital Media player
1	Samsung Blu Ray Media Player
6	Ceiling Mount for Projector
4	Phillips Media Player

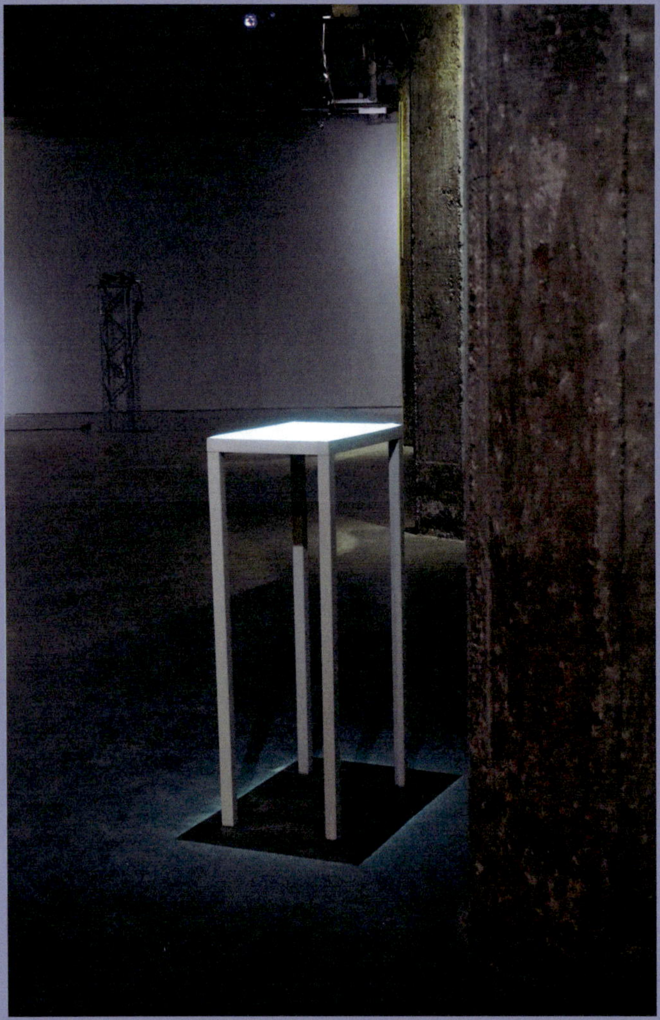

Installation view of Shilpa Gupta's *100 Hand drawn Maps of India.*

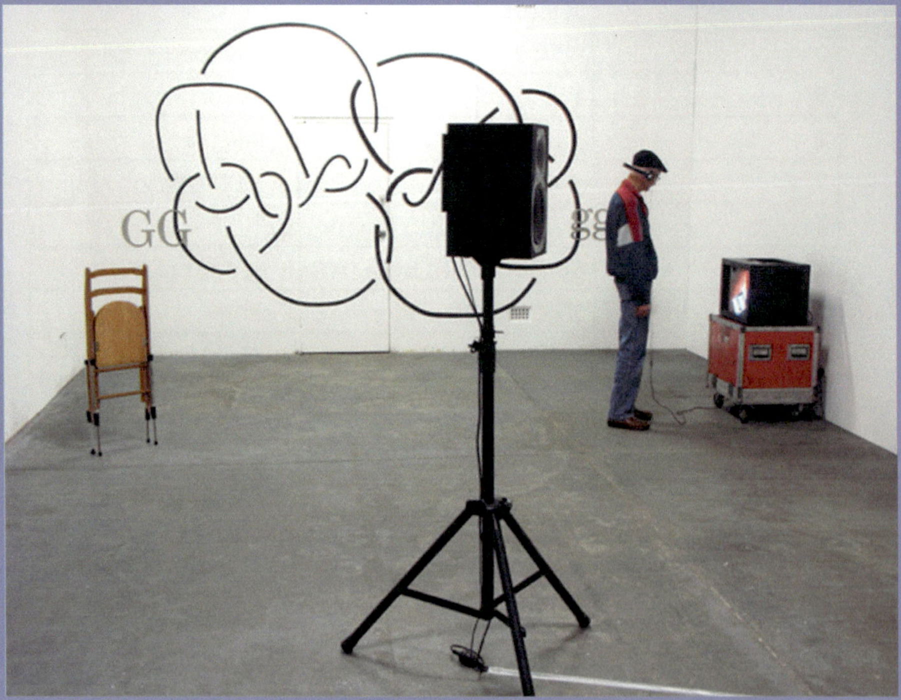

Installation view of Tim Woodward's *Concertistic LIfe* at Boxcopy Contemporary Art Space.

This year, MAAP Media Bank equipment has helped create artworks and sponsored artists presenting works featured in Sydney Art Month, Brisbane Experimental Arts Festival, The NAB Emerging Art Prize, and The Melbourne Art Fair. MAAP has also continued to support the exhibition programs of local ARIs such as OtherFilm and The Wandering Room; emerging galleries such as Boxcopy and The Hold Artspace; and established institutions such as the Institute of Modern Art, Spiro Grace Art Rooms, David Pestorius Projects, and Metro Arts.

In 2014, MAAP Media Bank also introduced an annual membership fee ($50.00 per artist). This membership fee enables artists to continue to borrow equipment as they require (subject to availability) while also helping to raise much needed funds for maintenance of equipment and payments to space and staff, while we seek additional support through funding grants and from the general community.

Feedback gathered at the completion of each loan process indicates that MAAP Media Bank continues to deliver a much-needed service to the artistic community. Without using the free audio-visual equipment through the Media Bank loan

Examples of feedback:

"It was fantastic to have MAAP support our big project for 2014, 'Institutional Capture'...The ability to access and utilize the extraordinary amount and type of gear that MAAP offer was invaluable - it enabled our artists to achieve what they wouldn't otherwise be able to achieve."

Dani Zuvela – 'Institutional Capture' by Otherfilm at Boggo Road Goal

"The artists would not have been able to present their new media works if we had not been able to loan the required equipment from MAAP. Consequently MAAP was invaluable in ensuring the artworks were shown as the artists' intended and the exhibition was executed as planned"

Sarah Barron – 'Compulsion Loop' at Boxcopy

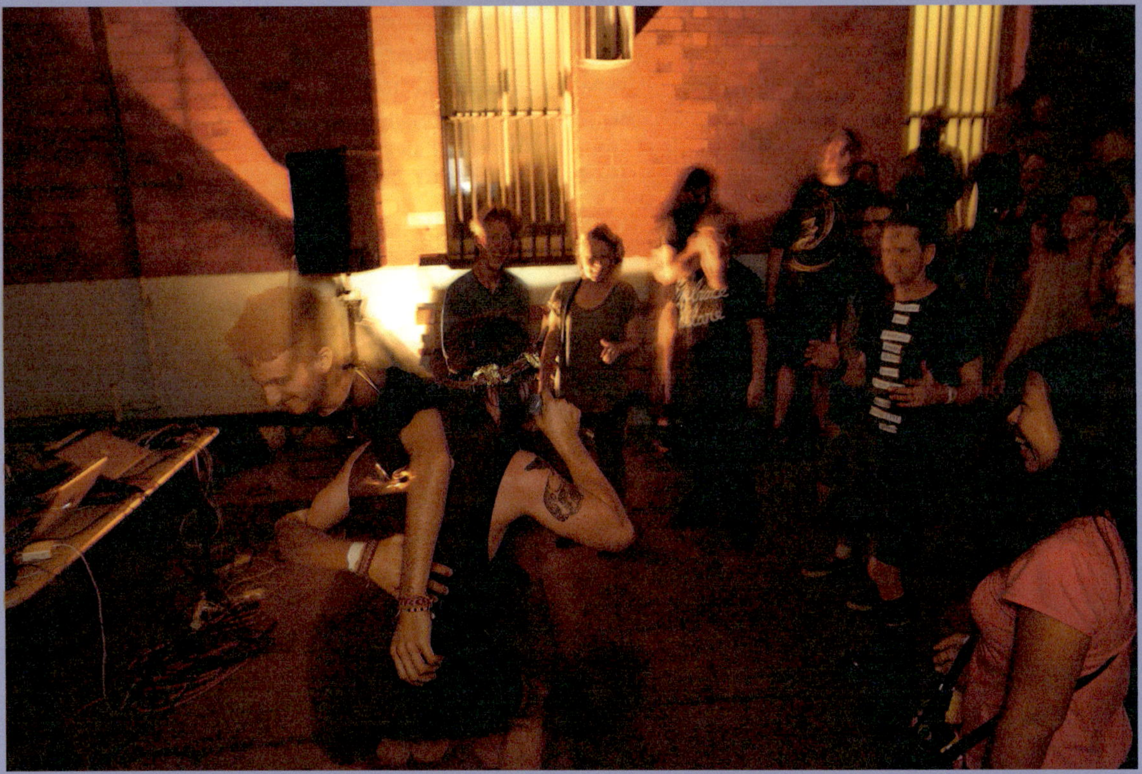

Above: OtherFilm event; Below: the opening of *Compulsion Loop* at Boxcopy Contemporary Art Space.

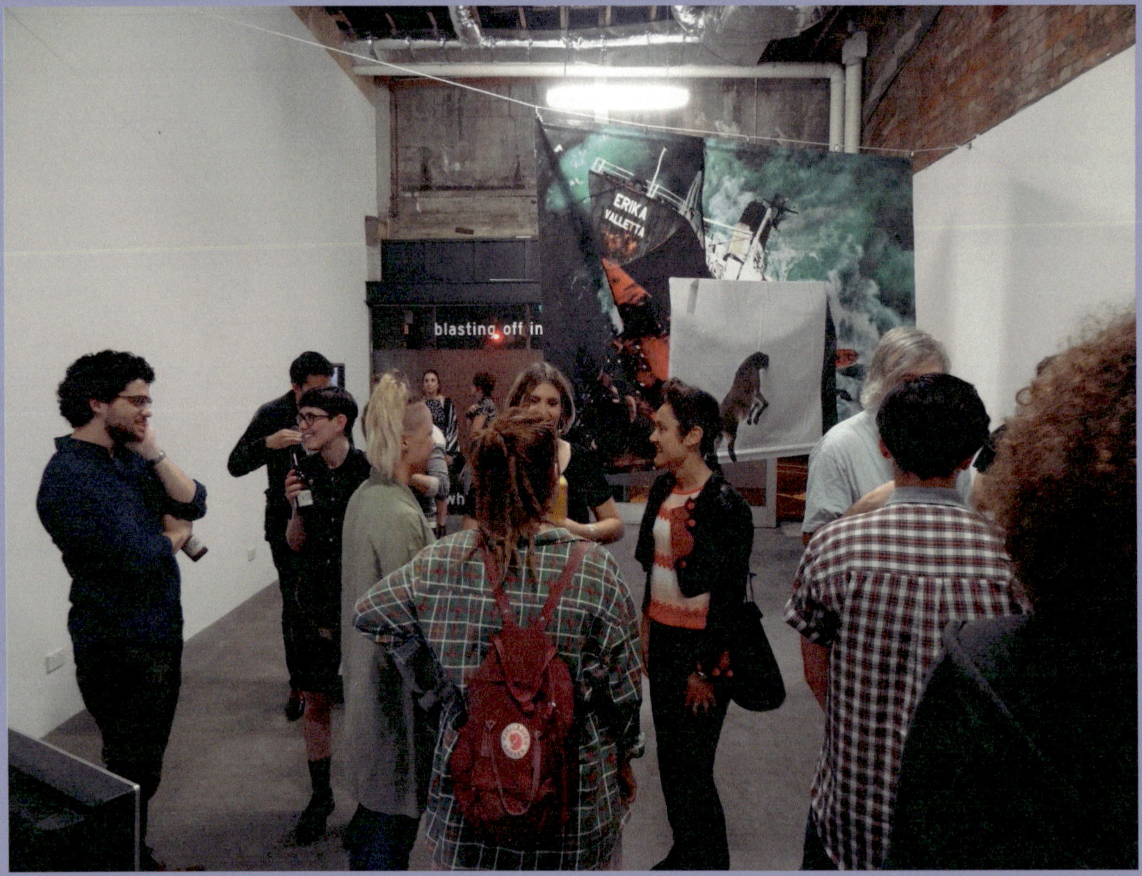

"In realising the exhibition Concertistic Life, MAAPs assistance was huge. The online Media Bank inventory enabled me to confidently plan the exhibition from interstate. Also, thanks to MAAP I was able to present my work at its best, using quality equipment that I wouldn't have been able to afford buying or hiring at an industry rate. This in-kind support is hugely valuable, and essential to artists working across digital media / sound / screen space. Hire fees can easily be the difference between an exhibition happening or not. MAAP Media Bank allowed for me to make the exhibition I wanted to make - without compromise - by providing quality equipment that did justice to the work. This is of course a good thing for artists and audiences alike.

My recent experience with MAAP was very positive and professional. And it was clear that MAAP does an incredible amount of work with limited staff and resources. At relatively short notice, I was able to loan all that I needed for the exhibition - with all equipment functioning properly, and no maintenance issues encountered."

Tim Woodward – 'Concertistic Life' at Boxcopy

"The MAAP Media Bank provided me with equipment necessary to complete work for the NAB Private Wealth Emerging Art Prize in Sydney. This is an invitation only prize held in association with Art Month Sydney and the 20/20 Event. 8 Artists have been selected to participate in the prize exhibition.

This is an important opportunity for me to show new work in a prestigious prize exhibition and in so doing access new audiences and networks for the expansion and development of my studio career.

I would not have been able to make the work for the show without the assistance of MAAP. I would have had to purchase or hire very expensive equipment, which I could not afford."

Camille Serisier – NAB Private Wealth Emerging Art Prize Finalist

The full details of loans and their values for 2014 is outlined in APPENDIX 1: MAAP Media Bank Loans Summary & Value.

4b.MAAP Consultancy: Queensland emerging artists, Mentorships, and Volunteers

Currently, MAAP operates with a core of three waged staff (2 FT positions) and regularly works with volunteers from arts audiences, artist run initiatives, tertiary students, postgraduate students, and employment programs.

The success is reflected in the ongoing flow of arts industry and interested individuals approaching MAAP to gain industry experience. For example in 2014 MAAP hosted an Art History student from the University Queensland fluent in Mandarin and particularly interested in Asian contemporary art – she volunteered in Brisbane and travelled to support projects in Shanghai and Beijing and is now committed to a career the arts. Another Art History student with Korean background, volunteered to work on MAAP's exhibition in South Korea gaining valuable industry experience while contributing her Korean language skills to the project. As an organisation, MAAP is mindful of each volunteer's skills and background and respectfully engages their assets, which results in productive outcomes for both host and participant.

Through employment programs in 2014 MAAP hosted:

- A Chartered Accountant (6 month placement) – worked on bookkeeping and accounts
- An Architect fluent in Chinese and Japanese with 15 years experience in a State Government design department (6 month placement) worked on drafting museum floor plans and modeling 2 and 3D exhibition plans.
- A Business Management graduate specialising in Marketing and Advertising (5 month placement) worked on marketing and branding assessment; web site content
- A Computer Engineer with Masters Degree (8 month placement) worked on IT issues in Media Bank and office.

The efficient and responsive use of volunteers strengthens MAAP, gives valuable experience and enriches our community.

5. REPORT CONCLUSION

MAAP has:

- assisted the development of new work,
- engaged emerging curators and artists in its programs,
- established local audiences for national and international content, and
- brought new international audiences to Australian content.

MAAP generated:

• hands-on approaches to presentation outcomes (collaboration with artists and curators to present work in the best manner possible through access to equipment and exhibition design),

• promotion of media art in Queensland and internationally, which has engaged new audiences and fed critical discourse, and

• a contribution to media art culture nationally/internationally feeding artistic vibrancy and support of media art growth in Queensland.

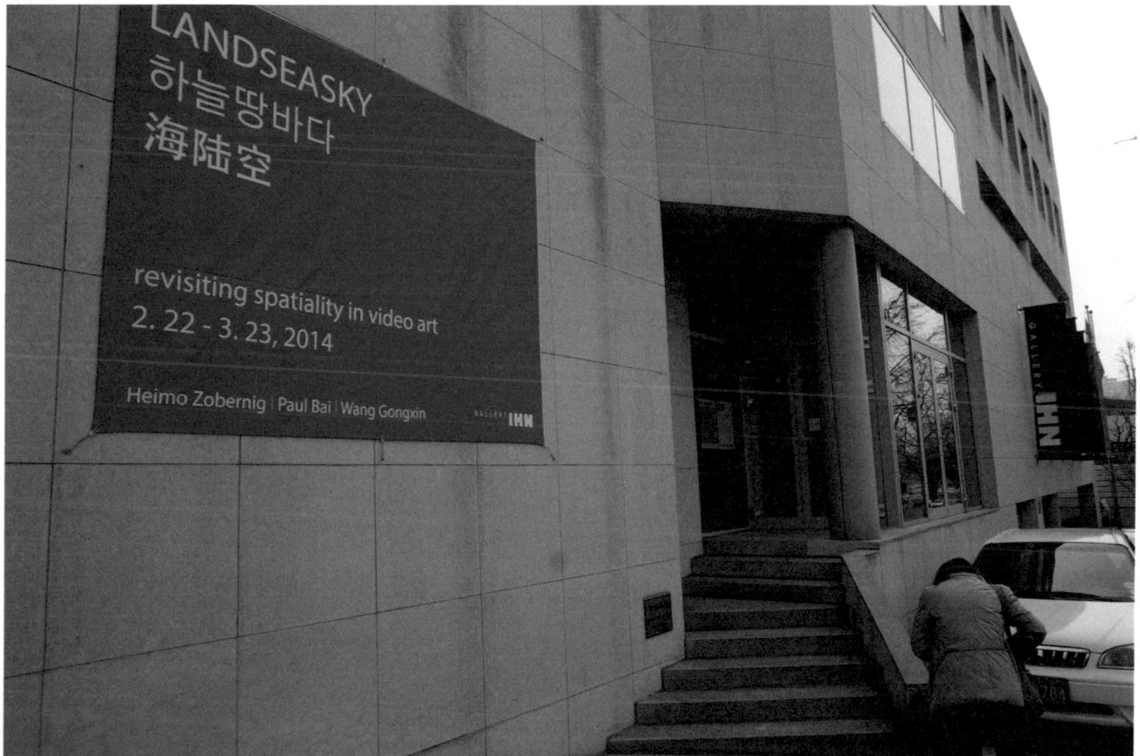

LANDSEASKY exhibition banner outside of Gallery IHN in Seoul.

APPENDIX 1: MAAP MEDIA BANK LOANS SUMMARY AND VALUE

MAAP Media Bank Loans Summary: 2006 – 2014

Total value of MAAP Media Bank Loans	**$1,495,967**
Total number of Projects Supported	**320**

Yearly Loan Breakdown

2006 Total Value of Loans	$8,000
2 Projects supported	
2007 Total Value of Loans	$43,320
13 Projects Supported	
2008 Total Value of Loans	$75,209
34 Projects Supported	
2009 Total Value of Loans	$404,674
57 Projects Supported	
2010 Total Value of Loans	$122,865
40 Projects Supported	
2011 Total Value of Loans	$243,335
39 Projects Supported	
2012 Total Value of Loans	$163,222
43 Projects Supported	
2013 Total Value of Loans	$106,609
51 Projects Supported	
2014 Total Value of Loans	$328,733
41 Projects Supported	

The following pages detail the MAAP Media Bank loans made in 2014.

2014 MAAP Media Bank Loans

Artist/s	Exhibition	City	Venue	Dates equipment hired	Equipment	Duration	Value
Grace Kevill-Davies, Kinlyside, Sarah Poulgrain, Tom Brooks	'Earth wind and fire'	Brisbane	Boxcopy Gallery, 282 Petrie Terrace, Red Hill	12/2/2014- 19/3/2014	1 x NEC Projector, 1 x DVD player, associated cables and consultation	35 days	$1,165
Athena Thebus, Chris Bennie, CLark Beaumont, Louise Bennett and Pirrin Francis	Presence/ absence	Brisbane	Spiro Grace Art Rooms	2/02/2014- 5/03/2014	2 x Sony Bravia 40" LCD Monitors, remote, power cables and consultation	31 days	$4,640
Bremen Town Musician	Album Launch	Brisbane	The Waiting Room, Browning St, West End	5/02/14	consultation	1 day	$70
Chenaya Bancroft-Davis, Aaron Butt, Daniel Cabrera, Amy Commins, Katherine Clayton, Romii Fulton-Smith, Claudia Greathead, Hillary Green, Jenna Green, Loki Groves, Alex McGovern, Sophie Richards	New World City	Brisbane	The Hold Artspace, Montague Rd, West End	12/02/2014 - 05/03/2014	1 x NEC 1500 Data Projector, 1 x 32" Sony Bravia Monitor, 2 x remote controls, 1 x Monitor wall mount, 2 x DVD players, associated cables and consultation	21 days	$2,495
Jarrod Van Der Ryken, Madeleine Stack	New World City	Brisbane	The Hold Artspace, Montague Rd, West End	12/02/2014 - 05/03/2014	1 x Digital video Camera, 2 x DVD players, 2 x 32" Sony LCD Monitors, associated cables and consultation	21 days	$5,480
Wilkins Hill - Solo Exhibition	Brainal Pipes Confusion Cave	Brisbane	Metro Arts, Brisbane CBD	12/02/2014 - 12/03/2014	1 x set of Edirol Speakers, 1 x 32" Sony Bravia LCD Monitor, remote, power cables and consultation	28 days	$2,300

Artist/s	Exhibition	City	Venue	Dates equipment hired	Equipment	Duration	Value
Jan Dibbets (The Netherlands), Paul Bai (Australia), Lauren Brincat (Australia), Barbara Campbell (Australia), Wang Gongxin (China), Shilpa Gupta (India), Yeondoo Jung (South Korea), Derek Kreckler (Australia), Giovanni Ozzola (Italy), Joao Vasco Paiva (Portugal/Hong Kong), Wang Peng (China), Kimsooja (South Korea), Craig Walsh (Australia), Sim Cheol Woong (South Korea), Heimo Zobernig (Austria), Yang Zhenzhong (China), Zhu Jia (China)	LANDSEASKY International Touring Exhibition	Seoul, Korea	Artsonje Center, Lee Hwaik Gallery, One and J Gallery, Opsis Art, Gallery IHN, Gallery Skape	15/1/14 - 15/7/14	5 x Mitsubishi Data Projectors, 5 x NEC Data Projectors, 5 x Sharp Data Projectors, 1 x BenQ Data Projector, 1 x Blu Ray Player, 6 x Mac Mini Computers, 1 x Projector Ceiling Mount, 1 x Computer Keyboard, 1 x Computer Mouse, 8 x WD Media Players, 8 x Brightsign Media Players, 5 x Phillips media Players, 1 x HD network Media Player, associated cables, remotes and consultation	180 days	$116630
Dereck Kreckler	Artwork Production	Sydney	N/A	08/01/2014 - 21/2/14	1 x Mitsubishi 4000 lumen DLPM Projector, associated cables, remote and consultation	44 days	$1,213
Amanda Wolf - Solo Exhibition	Peepshow/ Creepshow	Brisbane	The Hold Artspace, Montague Rd, West End	19/03/2014 - 02/04/2014	2 x NEC Data Projectors, 2 x Laser DVD players, associated remotes, cables and consultation	14 days	$800
Cigdem Aydemir, Karen Black, Jen Broadhurst, Athena Thebus , Bianca Beetson, Clark Beaumont, Kelly Doley, Tamsin Green, Chantal Fraser, Chayni Henry, Hannah Raisin, Tyza Stewart , Hillary Green, The Motel Sisters, Sancintya Simpson, Jodie Whalen	Common Woman	Brisbane	Level ARI at Metro Arts, Brisbane CBD	12/03/2014 - 16/04/2014	3 x Sony Bravia 32" Monitor, Monitor wall mount, associated cable, remote and consultation	35 days	$7,760

Artist/s	Exhibition	City	Venue	Dates equipment hired	Equipment	Duration	Value
A.D.S. Donaldson, Ant Farm, Amelia Barikin, Catherine Chevalier, Chris Wilder, Dave Hullfish Bailey, Diedrich Diedcrichsen, George Tjungurrayi, Janet Burchill + Jennifer Mccamley, Justin Clemens, Liam Gillick, Mark Fisher + Justin Barton, Mark Webb, Mary Webb, Melinda Harper, Peter Weir, Ralph Balson, Rex Butler, Sally Butler, Sam Watson, Timothy Dempsey Tjupurrula, Yukultji Napangati	Triple Oh!	Brisbane	David Pestorius Projects gallery	9/4/2014 - 27/09/2014	2 x Proel Black Speaker Stands, 2 x Behringer Truth Studio Monitors, 1 x NEC 2500 Lumen Projector, associated cables, remotes and consultation	171 days	$8,140
Paul Bai, Janet Burchill & Jennifer McCamley, A.D.S. Donaldson, Andreas Exner, Melinda Harper, Gary Warner, and Heimo Zobernig	Spring 1883	Melbourne	Hotel Windsor, Melbourne CBD	09/4/14 - 26/11/14	1 x Edirol Speaker Set, 1 x Mitsubishi 42" LCD Monitor, associated cables, remotes and consultation	232 days	$17,890
Sonja Payes as part of group exhibition	QLD Centre of Photography Exhibit	Brisbane	Percolator gallery, Paddington	16/04/2014 - 30/04/2014	1 x 40" Sony Bravia LCD Monitor, associated cables and consultation	14 days	$1,140
Tim Woodward - Solo Exhibition	Concertistic Life	Brisbane	Boxcopy Gallery, Red Hill	12/05/2014 - 11/06/2014	2 x Behringer Truth B2031A Studio Monitors, 2 x Proel Black Speaker Stands & top bracket, 1 x Sony Bravia 32" LCD Monitor, 1 x wall mount, associated cables, remote and consultation	30 days	$3,150
Megan Cope - Solo Exhibition	The Blacktism	Brisbane	Metro Arts, Brisbane CBD	15/05/2014 - 10/06/2014	1 x Projector, 1 x DVD player , 1 x Audio Interface, 4 x of headphones, associated remotes, cables and consultation	26 days	$1,040

Artist/s	Exhibition	City	Venue	Dates equipment hired	Equipment	Duration	Value
Jan Dibbets (The Netherlands), Paul Bai (Australia), Lauren Brincat (Australia), Barbara Campbell (Australia), Wang Gongxin (China), Shilpa Gupta (India), Yeondoo Jung (South Korea), Derek Kreckler (Australia), Giovanni Ozzola (Italy), Joao Vasco Paiva (Portugal/ Hong Kong), Wang Peng (China), Kimsooja (South Korea), Craig Walsh (Australia), Sim Cheol Woong (South Korea), Heimo Zobernig (Austria), Yang Zhenzhong (China), Zhu Jia (China)	LANDSEASKY Dispatches	Brisbane	MAAP Space	16/05/2014 - 07/10/2014	2 x NEC Data Projectors, 2 x WD Media Players, associated cables and remotes	144 days	$8,405
Camille Serisier	The Wonderful Land of Oz	Sydney	Art Month and NAB Emerging Art Prize	23/07/2014 - 25/07/2014	3 x ARRI 300plus fresnel spotlights with barn doors & stands and consultation	2 days	$680
Karike Ashworth - Solo Exhibition	Lamentation	Brisbane	The Hold Artspace, Montague Rd, West End	10/07/2014 - 24/07/2014	1 x Sony Bravia 40" Monitor, 1 x wall mount, 1 x Sony TV remote and consultation	14 days	$1,220
Zoe Knight, Grace Kevill-Davies, Amy Commins, Jamie Behrendorff, Courtney Pedersen, Sandra Selig, Ruth McConchie	Brisbane Experimental Art Festival (BEAF)	Brisbane	Judith Wright Centre for Contemporary Art, Fortitude Valley	04/08/2014 - 11/08/2014	1 x Sony Bravia 40" Monitor, 1 x wall mount, 1 x Sony TV remote and consultation	7 days	$5,180

Artist/s	Exhibition	City	Venue	Dates equipment hired	Equipment	Duration	Value
Jan Dibbets, Paul Bai, Lauren Brincat, Barbara Campbell, Wang Gongxin, Shilpa Gupta, Derek Kreckler, Joao Vasco Paiva, Kimsooja, Zhu Jia	LANDSEASKY International Touring Exhibition	Sydney	National Art School, Sydney	1/8/14 - 1/11/14	3 x Sharp Data Projectors, 7 x Sony Data Projectors, 1 x Mitsubishi Data Projector, 2 x Sony Headphones, 4 x DVD players, 1 x LG 14" Monitor, 7 x Projector Ceiling Mounts, 1 x Computer Mouse, 1 x Edirol, 1 x Phillips Media Player, 4 x WD Media Player, 3 x Brightsign media Players, 1 x Mac Mini Computer	93 days	$39,392
Jan Dibbets, Wang Peng, Zhu Jia, Yeondoo Jung, Paul Bai, Shilpa Gupta, Wang Gongxin, Heimo Zobering	LANDSEASKY International Touring Exhibition 1	Brisbane	Griffith University Art Gallery	09/09//14 - 21/10/14	6 x Phillips Media Players, 4 x NEC Data Projectors, 5 x Projector Ceiling Mounts, 5 x Mitsubishi Data Projectors, 1 x Mac Mini Computer, 1 x 15" Samsung Monitor, 5 x Brightsign Media Players, 3 x Sony Bravia 42" Monitors, 2 x Sharp Data Projectors, 1 x BenQ Data Projector, 1 x CRT TV	42 days	$29,470
Jan Dibbets, Barbara Campbell, Kimsooja, Derek Kreckler, Joao Vasco Paiva, Wang Peng, Zhu Jia	LANDSEASKY International Touring Exhibition 2	Brisbane	Griffith University Art Gallery	21/10/14 - 4/12/14	6 x Phillips Media Players, 4 x NEC Data Projectors, 5 x Projector Ceiling Mounts, 5 x Mitsubishi Data Projectors, 1 x Mac Mini Computer, 1 x 15" Samsung Monitor, 5 x Brightsign Media Players, 3 x Sony Bravia 42" Monitors, 2 x Sharp Data Projectors, 1 x BenQ Data Projector, 1 x CRT TV	44 days	$30,919

Artist/s	Exhibition	City	Venue	Dates equipment hired	Equipment	Duration	Value
Heimo Zobernig & Yang Zhenzhong	LANDSEASKY International Touring Exhibition 1	Brisbane	MAAP Space	1/10/14 - 5/11/14	2 x NEC Data Projectors, 2 x WD Media Players, associated cables	36 days	$2,377
Sim Cheol Woong	LANDSEASKY International Touring Exhibition 2	Brisbane	MAAP Space	5/11/14 - 28/11/14	3 x Panasonic Data Projectors, 3 x Mac Mini Computers, associated cables & remote	22 days	$2,152
Institute of Modern Art	Exhibition Support	Brisbane	IMA, Fortitude Valley	31/05/2014 - 09/07/2014	2 x NEC Data Projectors	39 days	$1,918
Sarah Poulgrain, Llewellyn Millhouse	Feel the Quality	Brisbane	Metro Arts, Brisbane CBD	1/09/14	consultation	1 day	$70
Camille Serisier	The Wonderful Land of Oz	Sydney	Art Month and NAB Emerging Art Prize	27/08/2014 - 03/09/2014	3 x ARRI 300plus fresnel spotlights with barn doors & stands and consultation	7 days	$1,870
Peter Alwast	The Origins Of Humanity	Brisbane	Boxcopy Gallery, Red Hill	10/9/14 - 24/09/2014	4 x behringer speakers, 4 x speaker stands, associated power cables and consultation	14 days	$780
Chris Bennie, Madeleine Stack, Marnie Edmiston	Compulsion Loop	Brisbane	Boxcopy Gallery, Red Hill	8/10/2014 - 15/10/14	3 x Sony Bravia 32" Monitor, 1 x monitor wall mount, 3 x Neumann headphones, associated cables, remote and consultation	7 days	$1,785
Amanda Wolf - Solo Exhibition	Touch Me I'm Sick	Melbourne	Strange Neighbour Gallery, Fitzroy, Victoria	2/10/14 - 22/10/14	1 x NEC Data Projector, associated cables & remote and consultation	20 days	$565

Artist/s	Exhibition	City	Venue	Dates equipment hired	Equipment	Duration	Value
Beau Allen, Alrey Batol, Karen Black, Peter Bruce Booth, Rohan Bridge, Aaron Butt, Natasha Cordasic, David Creed, Simon Degroot, Kristine Dragland, Nicholas Drake, Simone Eisler, Leah Emery, Simon Finn, Caitlin Franzmann, Bridie Gillman, Griffith University Art Gallery, Lily Halton, Tassia Joannides, Madeleine Kelly, Svenja Kratz, Danni McGrath, Dhana Merritt, Casselle Mountford, Jamie Mumford, David Nixon, Zoe Porter, Angela Rossitto, David Sawtell, Camille Serisier, Benjamin Sheppard, Kylie Spear, David Spooner, Lynden Stone, Gabriella Szablewska and Brent Wilson, Hiromi Tango, Rachael Wake, Libby Wallace, Manda Wolf and more	Tools of the Trade	Brisbane	Jugglers Art Space Inc, Fortitude Valley	15/10/14 - 5/11/14	3 x CRT TVs, 3 x DVD players, associated cables and remotes and consultation	21 days	$1,040
Joanna Bone, Peter Breen, Mel Davis, Joel Glazebrook, Paul Harris, Kay Lawrence, Matt Lockwood, Zoe Mary, Joy McDonald, Aaron Micallef, Rose Moxham, Jessica Row, Sha Sarwari, Vanessa Stanley, James Watts	The Northshore Hamilton Response to River Exhibition	Brisbane	The Shed, Hamilton Northshore	28/10/14 - 13/11/14	1 x Mitsubishi 4000 lumen DLPM Projector, 1 x NEC Data Projector, 3 x Sony Bravia 32" Monitors, 2 x DVD players, 4 x WD media Players, 2 x headphones, associated cables, remote and consultation	16 days	$4,993
Bonnie Hart	HR	Brisbane	Metro Arts, Brisbane CBD	5/11/14 - 12/11/14	3 x Neumann Headphones, consultation	7 days	$175
Institute of Modern Art	Exhibition Support	Brisbane	IMA, Fortitude Valley	11/10/14 - 29/11/14	2 x Panasonic Data Projectors, associated remotes and cables	49 days	$2,380
David Creed	Artwork Production	Brisbane	The Wandering Room,	22/10/14 - 3/12/14	1 x video camera, associated cables and consultation	42 days	$4,270

Artist/s	Exhibition	City	Venue	Dates equipment hired	Equipment	Duration	Value
OK YEAH COOL GREAT (Kate Beckingham & Anna McMahon), Kylie Banyard & Ron Adams, Monika Behrens & Rochelle Haley, Charles Dennington & Tully Arnot, SuperKaleidoscope (Kim Fasher & Sarah Mosca), Darren Seltmann & Vicky Browne, Todd Robinson & Mark Titmarsh.	Fleet	Brisbane	Metro Arts, Brisbane CBD	15/10/14 - 19/11/14	1 x NEC Data Projector, 4 x Neuuman Headphones, associated cables, remotes and consultation	35 days	$1,245
Nicola Morton	Otherfilm: Institutional Capture	Brisbane	Boggo Road Goal, South Brisbane	19/11/14 - 27/11/14	1 x Alcorn McBride Digital/Video Binloop, associated cables and consultation	8 days	$150
Andrew McLellan, Andrew Tuttle, Bonnie Hart, Caitlin Franzmann, David Spooner, Drew Daniel, Emile Zile, Gerald Keaney & Laura Hill, Kiah Reading, Leena Riethmuller, Rebecca Ross, Ruth McConchie	Otherfilm: Institutional Capture	Brisbane	Boggo Road Goal, South Brisbane	25/11/14 - 27/11/14	1 x NEC Short Throw projector, 2 x NEC Data Projectors, 3 x DVD players, 1 x Sony Bravia 32" Monitor, 2 x Beringher speakers, associated cables, remotes and consultation	2 days	$1,355
Camille Serisier	The Wonderful Land of Oz	Brisbane	Spiro Grace Art Rooms	5/11/14 - 26/11/14	4 x Neuuman Headphones, 1 x Logitech Speaker set, 1 x Edirol Speaker Set, 1 x Sony Bravia 40" Monitor, 1 x Wall Mount, 1 x NEC projector	22 days	$2,815

Artist/s	Exhibition	City	Venue	Dates equipment hired	Equipment	Duration	Value
Luke Roberts, Bianca Beetson, Nat Paton, Chris Doolan / Kjetil Andreas Knutsen, Robert Braiden, Harold Offeh, D.C. Curlew, Bob Saenz, Elizabeth Wymarra, Brent Dunner, Travis Beard, Annie Hogan	We Love Short Shorts Opening	Brisbane	Haven Espresso, Stafford Heights	26/11/14 - 10/12/14	1 x Mitsubishi Data Projector, 2 x Beringher Speakers, 2 x Speaker Stands, associated cables, remotes and consultation	14 days	$895
Luke Roberts, Bianca Beetson, Nat Paton, Chris Doolan / Kjetil Andreas Knutsen, Robert Braiden, Harold Offeh, D.C. Curlew, Bob Saenz, Elizabeth Wymarra, Brent Dunner, Travis Beard, Annie Hogan	We Love Short Shorts Exhibition	Brisbane	Haven Espresso, Stafford Heights	26/11/14 - 15/01/15	2 x 32" Sony Bravia Monitors, 5 x Neumann Headphones, associated cables, remotes and consultation	51 days	$1,720
Artists supported 92	Exhibitions supported 41	Number of days 1,642					
							TOTAL EQUIPMENT HIRE VALUE $321,733

Made in the USA
Monee, IL
05 November 2026

67380367R00033